STUDIES IN MODERN EUROPEAN LITERATURE AND THOUGHT

General Editors:
ERICH HELLER
Professor of German at Northwestern University

and

ANTHONY THORLBY
Lecturer in German
at the University of Sussex

LORCA

Also published in this Series

Ian W. Alexander: BERGSON
Günther Anders: KAFKA
Robert Baldick: THE GONCOURTS
Arturo Barea: UNAMUNO
E. K. Bennett: STEFAN GEORGE
W. H. Bruford: CHEKHOV
J. M. Cocking: PROUST
Arthur Cohen: BUBER
Wallace Fowlie: CLAUDEL
Hugh Garten: HAUPTMANN
Robert Gibson: MARTIN DU GARD
Marjorie Grene: HEIDEGGER
C. A. Hackett: RIMBAUD
Hanns Hammelmann: HOFMANNSTHAL
Geoffrey H. Hartman: MALRAUX
Rayner Heppenstall: LÉON BLOY
William W. Holdheim: BENJAMIN CONSTANT
H. E. Holthusen: RILKE
M. Jarrett-Kerr, C. R.: MAURIAC
P. Mansell Jones: BAUDELAIRE
P. Mansell Jones: VERHAEREN
Janko Lavrin: GONCHAROV
Janko Lavrin: LERMONTOV
Rob Lyle: MISTRAL
Richard March: KLEIST
José Farrater Mora: ORTEGY Y GASSET
Iris Murdoch: SARTRE
Renato Poggioli: ROZANOV
Theodore Redpath: TOLSTOY
L. S. Salzberger: HÖLDERLIN
Elizabeth Sewell: VALÉRY
Cecil Sprigge: CROCE
Enid Starkie: GIDE
J. P. Stern: ERNST JÜNGER
Anthony Thorlby: FLAUBERT
E. W. F. Tomlin: SIMONE WEIL
Martin Turnell: RIVIÈRE
Bernard Wall: MANZONI

Other titles are in preparation

LORCA

AN APPRECIATION OF HIS POETRY

BY

ROY CAMPBELL

BOWES & BOWES
LONDON

First published in 1952 in the Series
Studies in Modern European Literature and Thought
by Bowes & Bowes Publishers Limited

EDITOR'S NOTE

Considerations of space have made it necessary to omit from the present text all quotations in the original Spanish. Although this is a regrettable deviation from the practice adopted for this series, the quality of the substitute—Roy Campbell's rendering in English of Lorca's poetry—may make up for what accomplished readers of Spanish lose; and these will find it not too difficult to obtain the originals, the titles of which are given in the bibliography at the end of this volume.

AUTHOR'S NOTE

Acknowledgments are due to *The Catacomb*, *Nine*, *Poetry Review*, and *Forum* for permission to reprint some of the translations included in this book.

R. C.

DEDICATED

TO ROB LYLE

Published by
Bowes & Bowes Publishers Limited
42 Great Russell Street, London, W.C.1
and printed by photolitho by
Unwin Brothers Limited
Woking and London

Reprinted 1961

CONTENTS

I. The Regional Poet . . . . . . page 7

II. The Early Poems 25

III. The *Romancero Gitano* 40

IV. The *Canciones* 61

V. The Dramas 67

VI. *Cante Jondo, Poeta en Nueva York,* and *Llanto por Ignacio Sánchez Mejías* 70

Biographical Note 78

Lorca's Published Work 79

I

The Regional Poet

The time is past when Lorca's steadily growing popularity could be ascribed to the dramatic circumstances of his murder during the Spanish Civil War. The memory of the wave of publicity on which his name was first launched as a martyr for communism would by now have caused a reaction against him in English-speaking countries if the worth of his work had not transcended all political emotions. Even in literary circles, where sincerity is so often subordinated to book-sales, and where ideas, long due to be buried, are kept going by means of artificial respiration, such as that administered by the Third Programme of the B.B.C., we can see from such books as *The God that Failed* that authors who proved the most gullible to sensational propaganda during the events surrounding Lorca's death are belatedly trying to readjust themselves to the actual world—unfortunately some fifteen or twenty years after it really matters.

That Lorca's reputation has continued to grow is a sign that good taste still lurks furtively somewhere in the background, in spite of all the changeable political emotions rampaging in the limelight. It is well established that Lorca had no political tub to thump; that his murder was due to a private grudge; that it was perpetrated under the cover of the general epidemic of killings; and that his was but one of tens of thousands of deaths which were due to the settling of personal accounts behind the smoke-screen of civil strife.

The most formidable criticism levelled against Lorca today comes from the very quarter from which he was first launched as a martyr. It reflects disappointment at the impossibility of substantiating the former claim of 'martyrdom' from his work; a reaction which would be further aggravated by the realisation that Lorca was by birth a landowning 'kulak' and by faith a Catholic. But as far as England is concerned, no book about Lorca has so far appeared which has not tried to use him as a political symbol. Elsewhere Lorca's former admirers and present detractors show that they fully realise this. They tell us now that he has no 'world-message', that he is regional and provincial, 'and therefore a very minor poet, after all'.

But surely, to have a 'world-message' is the very sign of a poetaster in modern times. Every sub-poetaster is hot-gospelling world-messages at the rate of a machine gun firing blank rounds. These world-messages are all the same old stale H. G. Wells-cum-

Shavian story about world-federation, world-state, and 'cosmic man'—a tale that was anachronised years ago as soon as civil war became the norm of all present and (for many years to come) all future wars. It is surely a sign of superiority not to have any message at all—any more than Shakespeare or Homer had one. In art, it is the particular that is the only gateway to the universal. Who is more 'provincial' than Cézanne was, or his great poetic contemporary Mistral? They hardly noticed anything that happened outside of their 'Provincia Romana', or, as it is now called, Provence. Yet both are more universally loved than their Provençal friend, the world-gospelling Zola, who was always lecturing them. There was, however, a time when that same 'Province', by its very 'Provinciality', imposed its language on all the courts of Europe as the universal literary language of the civilized world, and when our own Kings such as Edward II and Richard I (two very fine poets) wrote in that language, better than in their own, poems that are readable today.

Though not by any means the greatest Spanish poet of his time, Lorca is the most intensely and nationally Spanish. This, however, he does not express in patriotism, as an Englishman would do were he so extremely English as Lorca is Spanish. He is also one of the most narrowly regional of Spanish poets; and at the same time, paradoxically enough, he is the most popular and universal in his appeal both inside and outside Spain. His appeal is never more universal than when he is writing, at home, about his native Andalusia. It is never more parochial and provincial than when he is self-consciously trying to be 'cosmopolitan', under the influence of Whitman, in the poems written in, and about, New York and the Caribbean. Andalusia is Lorca's *querencia*. The *querencia* is the exact spot which every Spanish fighting bull chooses to return to,. between his charges, in the arena. It is his invisible fortress or camp. It is not marked by anything but the bull's preference for it, and may be near the centre, near the barricade, or between the two, as the bull chooses. The nearer the bull is to his *querencia* or stamping-station, the more formidable he is, the more full of confidence, and the more difficult to lure abroad into the territory of the bullfighters, for *their* territory is wherever the bull is most vulnerable, and least sure of himself.

During Lorca's sojourns abroad, or in Madrid, he always returned for poetical strength to his native province; even when he did not return to it in person, he returned in imagination, memory, and dreams: and it never failed him as a source of strength and inspiration. It was from there that he was most difficult to lure abroad into the intellectual territory of the enemy, the territory of bad verse and of adverse criticism, where the critics can take advantage of the vulnerability of the poet, just as the 'toreros' do of the weakness of the bull when he stays outside of the magic and magnetic radius of his *querencia*. A poet can only get into that enemy territory by writing poorly about things he cannot feel instinctively. That is what happens to Lorca when he leaves Andalusia. He is like his ancient fellow-countryman, the giant Antaeus, who rose with strength redoubled each time he touched his mother-earth.

The cities of Granada, Córdoba and Sevilla, the three capital cities of Andalusia, always recur in that order in the poems of Lorca. After Granada the ancient Roman Córdoba comes second in his heart. He is attracted by something shadowy, nostalgic, and melancholy in both these towns which have outlived their greatest splendour. Sevilla, the gay, beautiful, and ever young giantess with the huge carnation between her teeth is too powerful and raucous to engage the same tender and intimate love that he feels for Granada or Córdoba.

That Lorca, one of the most narrowly regional poets of modern times, should be at the same time one of the most universally appreciated among his contemporaries, is just one more of those delightful little paradoxes which make bearable the present stereotyping dehumanisation, unification, and bureaucratic centralisation of human life. This universal appreciation is the mild, humble, and corrective protest of healthy human nature against the colossus-fetishism and political elephantiasis of the herd instinct which characterizes our time. It is a protest against the sacrifice of all real and immediate values to those ridiculous Utopian abstractions invented to glut that helpless and voracious credulity which is the inevitable substitute for lost faith.

It is precisely because Lorca has roots in his native soil that he can teach us what we need in spite of having no 'message'; and animate other European literatures and languages, in spite of the fact that he was ignorant of them.

* * *

Federico García Lorca, the son of landowning gentry, was born at Fuente Vaqueros near the city of Granada in January 1899. Disqualified by an early crippling disease from taking part actively in the strenuous equestrian life of the stock-raising, horse-breeding, and soil-tilling community, of which he was later to become the lyrical voice, he nevertheless absorbed his early environment through his eyes and ears, and dominated intellectually whatever he was unable to experience physically. He became the real voice of the Andalusian landscape, of its village and town life, its bull-ring, its cattle-pastures, its gipsies, and its Church. At the age of five he could strike terror into the farm hands and the servants by imitating in voice and gesture, and embellishing from his own inspirative imagination, the hell-fire sermons of the local priest, which he delivered with tears streaming down his face, to the awed consternation of his audience. Fully to appreciate the poetry of García Lorca we must consider how much the vocal element predominates over the printed letter in Spanish life; and Lorca's poems were known long before they were ever printed.

Throughout Spain, and especially in Andalusia, there is a living popular tradition in verse, music, and dance. Nearly all over Spain pedlars of ballads travel round, from farm to farm, to recite or sing their compositions, and sell their broad-sheets if their clients are literate, or if not, teach them the airs and words by heart. Conversation, oratory, and sermons in Spain are all on a far higher level than in countries where the printed word has entirely ousted the spoken word as a medium for ideas, in spite of the mechanical progress of radio. I have seen the huge Cathedral at Toledo thronged with every class of hearer, from the least literate to the most learned, from the most militant atheists to the most devout, to listen to a sermon by the Carmelite Father, the so-called 'Pearl of Castile', which lasted for an hour and a half after my own powers of concentration had given way. The interest by which this preacher held half of his audience was not so much religious, as the interest of an expert audience in the possibilities of the spoken word, organised rhetoric, and even the gestures and facial expression that accompanied the verbal delivery. They knew the rules of the game as an Englishman knows cricket. I was like a foreigner who did not know a wicket from a bat. I was a sophisticate of the printed word and this had precluded any sophistication in the art of spoken rhetoric. Four years as a talks-producer on the B.B.C. had only rendered me less able to appreciate speech in its natural, rhythmically spoken function. In order to comply with the mechanical requisites of broadcasting, and even more so with the auricular indolence of print-happy listeners, we talks-producers had to become the butchers of our mother-tongue, and the enemies of periods, sentences, euphony and rhythm. We were trained to mutilate periods into short snappy phrases, hacked by coughs, chuckles, exclamations, hesitations, and other dislocations of the normal flow of words. There is no doubt that the speeches of Demosthenes or Cicero, however readable in print, would have a soporific effect upon people who are so overliterate that their verbal sense is almost entirely visual. Nearly every man in England or the United States reads three times as many words as he speaks or listens to—every day—even on the advertisement hoardings while travelling to work. We are entirely satisfied if verse scans or rhymes to the eye on the printed page: and even when it is read aloud we are satisfied as to its form, since we automatically represent what we hear visually to ourselves so that while we are listening the words are printing themselves on the mind's eye.

Spain is the last European country where conversation remains a popular art; and where the peasants, though illiterate, are as a rule so very highly cultured that high society and intellectuals imitate their special accents and expressions, out of 'snobbism'—just as the new rulers of the bureaucracy affect and imitate the accents which are current at the moment in Oxford and Cambridge. As Bertrand Russell pointed out in *The Listener* a year or so ago, and as I pointed out in *Flowering Rifle* fourteen years ago, literacy is no guarantee of intelligence. It is often far more credulously vulnerable to exploitation, Utopian propaganda and any kind of advertising clap-trap than even the most primitive ignorance. The penniless kings of the sierra have an inherited culture founded on common sense and tested through centuries. This culture is far superior to the ready-made compulsory board-school education imposed on the rest of Europe which leaves its victim entirely at the mercy of the first demagogic charlatan who comes along with a few pseudo-scientific, political catchwords or psychoanalytical slogans. Nothing in the way of peasantry could be more different from the wretched revolution-fodder described by Tchekov, Zola, and Maupassant than the Spanish peasant,

> Who reads less nonsense from his running brooks
> Than writers primer-proud with knowing looks
> Can mumble out of newspapers and books.

We meet him through the ages (in the pages of *Quixote*, in the poems of Gabriel y Galán and in the plays of Lorca) with his absolute integrity, pride, and uncompromising courage.

When people speak about Spain being 'backward' they are judging her by a very false standard indeed. When progress was an uphill business, Spain was always very much up in the lead. When it became a veritable break-neck, downward, Gadarene stampede, Spain proved she was equipped with not only a brake in the form of tradition, but a reverse-gear in the form of reaction; and that as far as she was concerned, 'progress' was a matter of voluntary opinion. A body without reaction is a corpse: so is any social body without tradition. 'Reactionary' Spain has, during this century, produced better poetry than any other country; and this is chiefly due to her preoccupation with spiritual necessities rather than immediate physical conveniences.

At first sight some of the great poetical plays of Lorca will seem far-fetched because his characters prefer broken pockets, broken bank balances, and even broken hearts, to broken spirits. But those characters are true to life in Spain: and if they were true to life here, there would be more poetry here. There are no substitutes for morality, honour, and loyalty, either in themselves (as we are so painfully learning) or as the substance of poetry and drama. Their absence leads to boredom in life, and flat deadness in literature. There is no degree of poverty, hardship, or disease that is not preferable to the spiritual inertia which seems to be the price we pay for a certain amount of comfort and a bogus security; and that is the lesson that Spain and her poets have for us.

Spain is primarily a religious country; and one gets an instructive insight into the different temperaments of the different peoples comprising Spain by comparing how they observe their religion. In Castile the faith expresses itself intellectually in ideas, morals, devotion, mystical rapture, sanctity and heroism. In Andalusia it expresses itself aesthetically in colour, rhythm, and pageantry. As the Castilian language spreads southwards from its austere cradle of granite, ice, and fire on the high central plateau, which seems almost in reach of heaven, it softens its consonants and develops its vowels more voluptuously, just as a river, descending from snow-capped peaks and boulders to the plain, begins to sound less hoarsely and to move more musically and voluptuously when it comes to the groves of silver poplars, the olive orchards, and the low-lying acres of blossoming orange and lemon trees. It loses in impetus, strength, and crystalline clearness, to gain in colour, perfume, and breadth. The Castilian poetry of the high plateau is the supreme poetry of Spain: it searches the depths and heights of the heart, the intellect and the spirit. It is dynamic, whereas the Andalusian poetry is musical, descriptive, ornamental, pleasure-giving; and justifies its existence on those grounds rather than as the instrument of some compelling force or inspiration.

We have seen that Lorca's poetry first became known by recitation: his fellow students picked it up from him and repeated it so that it was well known and had had articles written about it, both inside and outside Spain, long before it was ever printed in book form. There is something in that which reminds us of the old tradition of the jongleurs and troubadours; of the early ballad literature of Spain, which is quite as rich as ours; and also of the great folk-epic of the Cid which was apparently handed down orally from minstrel to minstrel before it was finally copied out centuries later. The speaking of poetry is a far more exacting test of its quality than the visual reading of it in silence from a page. There is one poet in contemporary English literature who is deeply conscious, as Lorca was, of the *sound* of words; that is Dylan Thomas: he extracts the maximum of meaning from words through their sound. So does Lorca. They are both musicians: more than any poets of our times, they have studied the evocative force of sound in words. Thomas is a dealer in thunder and lightning, Lorca deals rather in the sound of rivers and leaves, and the irridescences of lights and waters. Strength is Dylan Thomas's salient quality, though he is by no means without subtlety. Subtlety is Lorca's chief quality though he is by no means without strength. They both derive from countries where there is a very strong musical and vocal tradition, and where singing is a natural function of the people. (The success of the Welsh miners' choirs in Spain was only equalled by that of the Spanish singers and dancers in Wales.)

In Andalusia a living popular tradition of song, music, verse and dance survives intact, so that cattlemen, muleteers and ploughmen can extemporise music and verse of a very high quality, as they work or ride through the fields. A very big proportion of the ordinary working population can play the guitar, one of the most difficult of all instruments. This popular tradition is one of the main influences in Lorca's work; but it has been vastly exaggerated by some critics, excusably perhaps, since it is so rare in the modern world that it immediately distinguishes a living modern poet who is lucky enough to have undergone that influence. But Lorca grafted on to the tree of popular tradition his own gorgeous ramification of the sophisticated and highly *literary* Gongorine tradition, which is the very opposite of the simple popular folk-lore with which he blended it so harmoniously. Thus he performed the miraculous operation of combining the most cultivated artifice of baroque poetry with the ingenuous art of the people, and of reconciling the treasures of the library to those of the earth and sky of his native sierra. He was influenced in turn by many of the other great poets of the Golden Age besides Góngora, such as Quevedo, Fray Luis de León, Saint

John of the Cross, Lope de Vega, Calderon de la Barca, Tirso de Molina, Herrera and others who formed a galaxy only equalled in literary history by that of our own Elizabethans and Jacobeans. (Unfortunately, poetry is so much more difficult to translate than prose that most English readers only know the great period of Spanish literature through its chief masterpiece *Don Quixote*, and they are unaware even that Cervantes was also a fine poet and wrote one of the greatest poetic dramas ever written, the *Numancia*.) Góngora probably influenced Lorca more than any other poet of the great age; and he seems to have loved him better and longer than any other.

Of all Spanish poets Góngora has been the most fantastically imitated, the most highly praised and the most violently condemned. There is no doubt that he is the most strikingly individual poet in Spanish literature, a symbolist who contains the whole of Mallarmé and a lot more besides, a symbolist three hundred years before his time. During his life stormy controversies raged about him: and ever since then there has never been a period in Spanish literature when he was not either in the blackest disgrace or in the full limelight of universal adulation. He has two distinct manners, one spontaneous and easy to understand as the songs of Herrick. But his other manner which gave rise to the term 'Gongorism' is the accentuation of all those characteristics that distinguish Baroque art. A super-stylised rhetoric is the result, in which far-fetched images and metaphors collide and splinter kaleidoscopically into the most dazzling patterns. To say he is obscure, as many have done, is an error; the involved, resounding, glittering, and hyperbolical rhetoric in which he indulges is undeniably a thing of the greatest brilliance rather than darkness. It is through the excessive glare of his brilliant light that he bewilders our vision:

> As when by whitest rays shot through,
> The sky, so bright that it is blue,
> Becomes so blue that it is black.

The second renaissance of Spanish literature, to which Lorca belongs, derives chiefly from the transplantation of French symbolism on to Spanish soil by the Nicaraguan poet Rubén Darío, greatest of the Spanish moderns. Although he was not an innovator he was an animator and a vitaliser. French is an economical, exact, precise, and almost niggardly language compared with Spanish. French symbolism was an extravagant growth, and, transplanted into the richer soil of the Spanish language, with its vaster vocabulary and melodic range, it grew and flourished so rapidly that in the last sixty years there have been more and better Spanish poets than for many generations before. Darío, Unamuno, Chocano, Antonio and Manuel Machado, Valle-Inclán, Gabriel y Galán, de Mesa, Jiménez, Salinas, Villalón, Vicente Aleixandre, Gerardo Diego, Dámaso Alonso, Cernuda, Guillén, Lorca, Alberti, Ridruejo, Bousoño, Valverde, the Paneros, and a dozen others, have made Spanish poetry pre-eminent in the modern world. With the exception of the almost primitive throwback Gabriel y Galán, all these poets bear the mark of the impact of French symbolism. So it is small wonder that the archetype of all symbolism, Góngora, came back into favour with the last two generations as did also that other great symbolist Saint John of the Cross.

Of Góngora, Lorca says: 'He is as naturally Spanish as Lope. In his final most characteristic work, he rejects the mediaeval tradition to seek the glorious old Latin tradition. He seeks in the air of his native Córdoba the voices of Seneca and Lucan. He models Castilian verses by the cool light of the Roman lamp; and raises to its greatest height the type of art which is pre-eminently and uniquely Spanish, the Baroque.'

The two Córdobese Romans anticipated in a curious way the art that was to flower in Andalusia, and in Córdoba in particular, not only in the Arabic times, but later with Góngora and Herrera, and even today in the popular gipsy song and music. Both Seneca and Lucan are severe yet ornate, and they often 'Gongorise' with far-fetched imagery so that we sometimes get a triplicate image in their work. This is faulty from a coldly classical point of view, but it has its compensations, since when it 'comes off' the effect is as beautiful as that of a triple rainbow.

Lucan and Seneca certainly anticipated the form of poetry that was to flower in their native town under the Moorish conquest, and eventually to influence Lorca not only through its effect on popular tradition but through first-hand study of the Moorish poets as collected and translated by Emilio García Gómez in his *Arab-Andalusian Poems*. In these poems again the image and the metaphor are all important, but the extravagance has been tempered by brevity. The Andalusian Arab poets of the eleventh century perfected a type of poem called the 'casida' which was a cross between a lyric and an epigram: in conciseness, freshness, and beauty, it rivals some of the finest gems in the Greek anthology. Casidas differ from the classical epigrams, however, in being more daring and fanciful. They are the most paradoxical blend of extravagance with economy. Introduced suddenly from the starkness of the African desert which sharpens the senses with a sort of visual and olfactory hunger and thirst, these Arab poets seem to have revelled in the luxuriance and grandeur of Roman Andalusia. Although they enjoyed the new feast of colour and fragrance which they acquired from their conquest, something of the old ascetism of desert life remained with them to discipline the riot of their fantasy in these rich new surroundings, and to refine their newborn extravagance with the hereditary economy of the desert-horseman and the camel-rider. So in their casidas a single image often suffices for a single poem. The result is the compression and concentration of great force within narrow limits, as in a cartridge—if one can imagine a cartridge whose explosive powder consists of pollen, rather than sulphur, saltpetre, and carbon. The first line of their casida acts as a sort of hair-trigger, and releases, simultaneously, a flash of gorgeous colour and a delicious mental perfume. Casidas might be described as miniature Gongorisms. Though fragmentary in size, they constitute wholes in themselves, just as a beautiful feather or a flower is a whole, though only the part of a bird or a shrub. To translate these casidas is like trying to handle one of those gorgeous silk-shot butterflies upon which one is afraid to breathe for fear of dispersing the miniature aurora-boreales of conflicting sheens into a mere puff of grey powder. It is as risky as trying to pick a newly blown hibiscus whose pistil, overcharged with golden powder, threatens to shed it into thin air if one's hand so much as trembles.

The Moorish influence made itself felt very deeply on Góngora, even if not from first-hand reading, yet through the vast scholarship of his own contemporaries, such as his adversary Quevedo, twenty years his junior, who could read Hebrew, Arabic, Greek, and Latin, and whose great sonnet on Lisi's golden hair is a typical example of that explosive extravagance concentrated into cartridge form which we find in the Moorish casidas. In this sonnet Quevedo almost out-Gongorises Góngora, and though the allusions are classical, the general effect is that of the hyperbole common to all Córdobese versifiers from Seneca down to the present day gipsies, who will describe a bull with long, dangerous horns as having 'horns like Cathedrals'. Nevertheless the brevity of this sonnet contrasting with its ornateness is truly Moorish:

> When you shake free your hair from all controlling,
> Such thirst of beauty quickens my desire
> That on its surge in red tornadoes rolling
> My heart goes surfing over waves of fire.

> Leander, who for love the tempest dares,
> It lets a sea of flames its life consume;
> Icarus, at a sun whose rays are hairs,
> Ignites its wings and glories in its doom.
> Firing its hopes (whose deaths I mourn) it strives
> To fan their ash into fresh phoenix-lives
> That, dying of delight, new hopes embolden.
> A miser rich, the crime and fate it measures
> Of Midas, starved and mocked with stacks of treasures,
> Or Tantalus—with streams that shone as golden.

If we deny the direct influence of the various schools, Latin, Moorish, and Gothic, upon each other, we are simply left with the extremely unlikely thesis that the mere climate and surroundings of Andalusia (and its intellectual capital, Córdoba) produced precisely the same effects at intervals of many centuries, on three such different races and languages as that of the Roman, the Arab, and the Spaniard, and imposed on them all that far-fetched imagery which reaches its climax in Góngora. Here is a fragment of the latter's *Soledad*, which almost forms a casida in itself for sheer figurative brilliance. All it wishes to say is that the sun was entering the sign of the Bull:

> The year its flowery station reached, and now
> Europa's robber, in a shape that lied,
> (A crescent moon the weapons of his brow,
> The Sun the shining bristles of his hide)
> Refulgent pride of heaven, as he blazed,
> On sapphire fields the gold star-clover grazed.

Again and again we come upon echoes of extreme Gongorism in Lorca—especially in such far-fetched images as when a child, seeing the full-moon reflected in the water, invites it to clash its cymbals in the comic *Ballad of Don Pedro*. But there is no image in Góngora that haunts Lorca more than that third line of the above extract:

> Media luna las armas de su frente
> ('A crescent moon the weapons of his brow')

which is echoed at least a dozen times in different ways in his verse. He sees even the boys who are bathing in the evening as being charged by the waves which reflect the crescent, as by cattle with lowered horns:

> Dense oxen of the waters charge,
> With lowered heads, the youngsters bold
> Who bathe between their crescent moons
> And undulating horns of gold.

and again in the same poem, the *Romance del Emplazado*,

> It will be in the night, the darkness,
> By the magnetic mountain streams
> Where the oxen of the water
> Drink up the rushes in their dreams.

Once more the image is modified in the *Muerte de Antonito el Camborio*—this time with a memory of the 'rejoneador' in the bullring, a horseman who fights with a javelin:

> When in the grey bull of the water
> stars strike their javelins.

The earliest reference that one can remember to the moon as the horns of a bull comes in Moschus. It is repeated beautifully in Góngora, Lope, and again in Salvador Rueda, the nineteenth century Andalusian predecessor of Lorca, who speaks of the mountains like bulls 'raising on their impressive heads the sharp-pointed crescent moon'; but for Lorca it has an even deeper significance, since in his childhood, he tells us, he had often heard the phrase an 'ox of water' from the peasants, who so describe a heavily-moving, deep watercourse 'to indicate its volume, weight, force, and momentum'. Lorca says, 'I have also heard a farmer of Granada say, "Willows love to grow on the tongue of the river". "Tongue of the river" and "ox of water" are images of the common people, yet they are very closely related to Góngora's way of seeing things.' The extremes of erudite and popular metaphor are seen here to meet in one. Visual image and metaphor are the very idiom in which Lorca thinks and speaks, and they form the unit of all the creations of Lorca from his almost nonsensical but very beautiful embellishments of lullabies, nursery rhymes, and popular refrains, which have no other justification than the charm of their sound and the pictures they embody, to the deeply tragic and profound poetry of his best plays.

Verlaine, Christina Rossetti, and in our day, Edith Sitwell have all exploited the evocative power of the nursery rhyme and the popular song. Shakespeare did it with terrific effect in *King*

Lear. Lorca has written a long essay on Spanish folk-lullabies, and he has made weird and haunting poetry out of them, as in the case of the lullaby at the beginning of *Blood Wedding*, which though it means little enough yet suggests the terror and tragedy lurking behind the first scene, in which it is sung by the Mother and the Grandmother, turn about, to the sleeping child:

> Sing lullaby, baby,
> About the great horse
> Who would not drink water,
> So black was its course . . .
> Who will say, baby,
> What the water is reaping
> Through the green halls
> With his long tail sweeping?
> His mane is frozen,
> His hurt hoofs stagger,
> Between his eyes
> Is a silver dagger . . .
> They went to the river
> And there in the flood
> Stronger than water
> The stream ran blood . . .
> He would not touch it
> From the wet bank,
> Silver with flies,
> His underlip shrank . . .
> To the hard mountains
> He whinnied remote
> While the dead river
> Flowed over his throat.
> Ah! the great horse
> That dreaded the flow!
> Horse of the daybreak!
> Grief of the snow!
> Wait there. Do not enter.
> Shade the window from beams
> With dreams of branches
> And branches of dreams.
> Wait there! Do not enter!
> To the mountains repair
> And in the grey valleys
> You'll meet with the mare . . .

In spite of its lack of meaning, this 'nonsense-rhyme' creates the same ominous atmosphere as the nonsense of Edgar in *Lear*, or the line spoken apropos of nothing in the same play—'Childe Roland to the dark tower came'—which gives one a grue down the spine for no apparent reason. Again and again we find Lorca going beyond the verge of meaning, but never irrelevantly, in order to create an atmosphere of sorrow or foreboding: or else of gaiety, as in the delightful semi-gibberish of his surrealist description of the fair in the gipsy city of Jerez de la Frontera before the police-raid, in the *Romance de la Guardia Civil Española*:

> O city of the gipsy people!
> Flags at the corners of the streets.
> With calabashes and the moon
> And cherries candied into sweets.
> O city of the gipsy people!
> Who can forget you who has seen?
> City of sorrow and of musk
> With towers of cinnamon between.
> When the night-time has arrived,
> The night-time of the night,
> Gipsy folk upon their anvils
> Are forging suns and darts of light.
> A wounded horse arrives and runs
> To all the doors with plaintive whine.
> Cocks of glass are crowing loud
> At Jerez of the Frontier-Line.
> Around the corner of surprise
> The wind bursts naked on the sight,
> In the night, the silver night-time,
> In the night-time of the night.

This is entirely playful in its parody of the nonsensical popular refrain 'cuando la luna lunera' (When the moon, the moony moon) yet it exactly conjures up the hyperbolised scene of any happy-go-lucky gipsy quarter in the South of Spain with the horse-shoeing going on after dusk, the wounded horse returning, and the wind being surprised without 'its shift'.

We have stressed the ornamental side of Andalusian literature as it appears in Lorca and the poets that have influenced him; but we have not spoken yet of the quality which tempers and balances it in the best of this literature—the quality of architecture and design which underlies and sustains its luxuriance just as the stoical geometrical Roman ruins underlie and sustain the more ornate synagogues and mosques of Córdoba, Granada, and Sevilla, with the huge masses of jasmine, bougainvillea, passion-flowers, and ivy which entwine them. Nearly all Lorca's most flowery work is vertebrated with a sinewy spinal-cord. However much he luxuriates externally into such sparkling froth as the above passage, it is the froth of good strong champagne—with a body to it. There is always a central design to his work. The two principles, that of luxuriation and that of economy, wrestle harmoniously in nearly all the best Andalusian art, as if a sort of tropical oriental luxuriance were being pruned and ordered by the cold hard axe of Romanity.

Andalusian art reached the summit of its expression only two years ago in the style of the bullfighter Manuel Rodríguez (Manolete) whom Federico García Lorca never had the good luck to behold. He was the apotheosis of Andalusian poetry, sculpture, choreography, and harmony. He was ornate yet severe, like Seneca. He was surnamed 'the tower of Córdoba'. He personified opposite extremes, at one and the same moment contrasting, in one being, the austerity, coldness, and hardness of a statue with the softness and warmth of a flower: for while his brain, his muscles, and his sword were dedicated to the clear and icy geometry of death, yet with his crimson cape and his scarlet muleta he sculptured, to the imagination, not only the shape and the colour, but, intellectually, even the perfume of the full-blown corollas of magnolias, roses, and camellias.

When Lorca is at his best, these two opposing principles balance each other in his work: and that he is fully conscious of the value of the discipline of the Roman bonework beneath the richer and softer clothing and flesh, he is always reminding us in memorable passages, as in his portrait of another great Andalusian bullfighter, Ignacio Sánchez Mejías, who, though not quite as great as Manolete, or Joselito, died, like them, in the bullring. Mejías was lamented by Lorca in the best dirge of all the hundreds that have been written about bullfighters killed in action, *Llanto por Ignacio Sánchez Mejías*. In this poem he describes perfectly how the Roman principle of proportion and design tempers the wild strength and extravagance of the Andalusian:

> Like a torrent of lions, his
> Incomparable strength was rolled,
> And like a torso hewn in marble
> His prudence carven and controlled.
> Gold airs of Andalusian Rome
> Circled his head and gilded it,

> Whereon his laugh was like a lily
> Of clear intelligence and wit.

There is also this grimly rectilinear image which serves as a reminder of the Roman gravity at the core of Lorca's inspiration. This comes at the end of the *Ballad of the Man who was Summoned to Appear before God* (*Romance del Emplazado*) and it describes how the white shroud settles over the body of the unhappy man:

> And with a hard clear Roman accent
> The spotless sheet around him rolled,
> Gave equilibrium to Death
> With the straight creases of its fold.

Then there are passages in which he burlesques Roman discipline and rectangularity but in such a way as to show how deeply it was inbred or inherited in him. There is the light-hearted passage in the ballad about Don Luis de Góngora riding out on his horse in which the wind, 'sculptor of bulky effigies', makes Roman sculpture of his flying cape: and the comic group of horsetail-crested and helmeted Roman soldiers in the *Romance of the Martyrdom of Olalla*. (The 'long-tailed horse' here is obviously the river splashing through the streets of the old Roman city of Merida):

> A long-tailed horse along the street
> Careers and leaps in foam
> Where idly doze or gamble
> The veterans of Rome.
> Half a mountain of Minervas
> Its leafless arms extends.
> Gilding the ridges of the rocks
> The hanging stream descends.
> Night of reclining torsos
> And stars with noses bust
> Waited for crevices of dawn
> To crumble into dust.
> Red-crested blasphemies
> From time to time resound.
> With her screams the young saint splinters
> The wine cups to the ground.
> Sharp hooks and knives upon the wheel
> Are honed with rasping sound.
> The bull of anvils bellows

> And Merida is crowned
> With half-awakened tuberoses
> And brambles all around.

In few passages in literature has such a clear impression been given of a complicated scene by means of so much transposition, inversion, litotes, hyperbole, and suggestion. It required an iron discipline to dominate it. We see the groups of helmeted soldiers, the camp-fires, mischievously lighting up the flattened noses of tough old veterans, and in the lines:

> Red-crested blasphemies
> From time to time resound

we can literally see, from time to time, a centurion rising in the light of a camp-fire with his red-plumed helmet and issuing an order with curses. We hear the screams of the doomed martyr so shrill that they crack the wineglasses (a favourite image with Lorca—as several famous Flamenco singers were renowned for breaking glasses with their voices). The instruments of torture being prepared in the smithy clang and rasp in the background: and the day breaks. As soon as they are read, the purposely confused images reassert themselves into a definite order and a picture which is almost Roman in its clearness: and we feel that Lorca selected a Roman image on purpose in order to delineate it so clearly by such completely unorthodox and apparently chaotic means. In this passage, if anywhere in Lorca, an exact discipline was operating though he chose the most recalcitrant material on which to impose it. The poem however is inferior to the great Latin poem of Prudentius on the same subject from which it is taken. The subject is really too tragic for the easy-going impersonal humour with which Lorca surrounds it.

We have seen that the image or metaphor (generally double or triplicate) is the unit of the Lorca idiom, and that at its best it depends on the fusion of more than one of the five senses into one another:[1] take the powerful image comprising the last two lines

---

[1] Continental critics are in the habit of tracing the birth of 'modern' poetry to Baudelaire's famous sonnet 'Correspondances', where, they say, for the first time in poetic history, the different senses were consciously expressed in terms of one another, scent in terms of sound, colour and sounds in terms of light and shade, and all in terms of each other. Professor Marcel Raymond in his book *From Baudelaire to the Surrealists* traces the modern metaphor to this sonnet. But surely the first man to make these 'correspondances' was the first man who ever

of the following quatrain from *Reyerta* or *The Brawl*. There has been a murderous brawl between gipsies, and the police are visiting the scene of the tragedy, where blood continues to trickle along the ground after being trodden and slipped on:

> The Judge and Civil Guard their way
> Along the olive orchard take,
> *Where slithered blood begins to moan*
> *the dumb song of an injured snake.*

Here, the eye and the ear are unaccountably coerced into a faultless union. The pain, the colour, the winding movement of spilt blood, and even the weakness resulting from it, are unforgettably printed on the mind.

Lorca's achievement, throughout his career, is reflected, as by an infallible index, by the effectiveness of his images. His poems stand or fall by them. They are the pulses by which we measure his health, beating strongly and normally in his *First Poems*, his best plays, the *Lament for the Bullfighter*, the *Canciones*, the *Cante Jondo*, but by fits and starts, irregularly, in his *New York Poems* and his *Diván de Tamarit*. Whereas a Castilian poet like Gabriel y Galán can write a superb poem without a single metaphor, simile or image, they form both the glory and (sometimes) the weakness of Lorca's work. He has periods when they are consistently successful, other periods when they are not, and it is very rare that he ever mixes good and bad images in the same poem.

spoke of 'sweet' sounds or music, thereby making a contact between the sense of taste and that of hearing—and he must have lived very long ago. In English, long before Baudelaire, we have had several highly sophisticated and conscious attempts to express sound in terms of light, perfume and colour. Wordsworth's superb image, 'beauty born of murmuring sound will pass into her face', surpasses Valéry, *in his own manner*, a century before he wrote! If Baudelaire did not know *Music's Duel* by Crashaw or Smart's *Song to David*, he must have known Shelley's famous *Ode to a Skylark* which copies the same formula as Crashaw uses in describing the voice of a nightingale. Shelley, exploiting the 'correspondances' far more thoroughly than ever Baudelaire did in one poem, triumphantly compares the voice of a skylark to the light of a glow-worm, the brilliance of a sunlit cloud, the radiance of a star, the scent of a rose, and many other things besides. The whole seed of the modern symbolist movement can be traced back to England: just as much as the French impressionist movement in painting can be traced back to Turner and Constable. If we cannot claim to compete in the finals, we may as well be content with having taken part in the preliminary heats.

II

The Early Poems

In the Collected Edition of Lorca's Works, printed as a prefatory introduction to his *First Poems*, we find a quotation from Lorca's spoken conversation with his friend and contemporary, the very excellent poet Gerardo Diego. It is revealing, not of any preconceived and conscious system by which Lorca worked, but of his total freedom from any such system:

> But what am I going to say about poetry? What can I say about these clouds, about this sky? To gaze, to gaze, to gaze at them, to gaze at it, and nothing more. Understand that a poet cannot say anything about poetry—that must be left to critics and professsors. But neither you nor I, nor any poet, knows what poetry is.
>
> Here you are: listen. I have fire in my hands. I understand it and work perfectly with it, but I can't speak about it without literature. I understand all poetics; I could speak about them if I did not change my mind about them every five minutes. I don't know. One day bad poetry may please me very much, just as today bad music pleases me (and us) almost to madness. I would burn the Parthenon tonight just to begin to rebuild it again tomorrow, and never finish it. In my lectures I've sometimes spoken of poetry, but the only poetry I can't speak about is my own. And not because I'm unconscious of what I'm doing. On the contrary, if it's true that I'm a poet by the grace of God—or of the devil—it's because I'm also a poet by the grace of technique and effort.

In this reported speech, which rings very true, we are reminded of the instinctive opportunism of Apollinaire, who also wrote like Lorca, straightforwardly, with grammatical structure, in the vernacular, and to whom all things whether beautiful, ugly, comic, or repugnant were potential subjects for poetry. Both wrote in simple language bordering on plain everyday speech—the antithesis to 'Hopkinese' or to Dylan Thomas's highly stylised personal language. They saw whatever was commonplace or quotidian with the fresh eyes of children. Sophistication and experience never exhausted that gift of wonder and genuine interest that transformed for the townsman and soldier, Apollinaire, the cavalry defences outside his trenches, or the lights on the tramlines in the suburbs, into the magical substance of poetry; or for the countryman in Lorca the commonplace objects of his landscape such as sisal into 'petrified octopus', or prickly pears into 'multiple batsmen' and 'savage Laocoöns'.

This power of enthusiastic perception is merely a matter of vitality—it is not an adopted attitude. In calling Lorca a 'nature poet' I mean something very different from what is meant in England. The influence of the natural landscape on a romantic northern poet generally, not always, results in a 'brown study', a blending of the intellectual attributes of the poet with the rocks, trees, and mountains around him, until they seem to become more articulate, reasonable, intelligent and sentient than the poet himself. There is a sort of 'merging' and self-annihilation in which the un-intelligent part of the landscape gets the better of the intelligent part of it, the poet, and finally swallows him. The poet disappears completely, into an inhuman mist which gets thicker and thicker, undisturbed by his dissolution, and apparently impenetrable to any supernatural rays from above. This operation is commonly practised by German, Scandinavian, and Anglo-Saxon poets; and it is judged to be highly laudable to be able to extinguish oneself in this way, as I know from far better poets than myself, though I have never been able to see the point of this self-immolation to an insensate universe. As a Celt and a Latin, nature leaves me cold, except as something to be dominated, confined, made to fructify, and loved, if at all, for the sake of the benefits it confers when it is properly treated, kept in its place, and rationed with water and manure. I like to see the sierra terraced from top to bottom by the skill and strength of men's hands, not lying uncultivated and wasting. I like to see a horse brought to the fullness of its strength and expression by having a skilled rider on its back, not running about as a brumbie trying to fend for itself on the backveld.

That is where Latin 'nature poets' differ entirely from the German and English Romantics. When I read Wordsworth, I feel that all his rocks, trees, and mountains are more intelligent than human beings, whereas his human characters, wherever they can be unearthed, are almost without exception imbeciles and nit-wits. I feel that there is something deeply perverse, intellectually suicidal and misanthropic about this transposition of values. In fact I think the whole Romantic principle (in its bad sense) can be defined as something perverse, a sort of centrifugal panic in which the poet escapes from himself: a principle that subordinates the immediate to the remote, the evident to the occult, the normal to the abnormal, the lucid to the obscure, the moral to the immoral, and the present to some Utopianised future or romanticised past. It necessitates many, or most, of these falsifications to credit a mountain, or a cloud, with having a quarter of the intelligence of an ant, let alone of a man. Aldous Huxley says that Wordsworth would never have written his poems on the benevolent intelligence of nature, if he had lived in the tropics amongst cobras, tarantulas, and scorpions. But Huxley commits as sentimental a perversion in seeming to ascribe a malignant intelligence to tropical 'nature'. I had two years as a jungle coast-watcher, almost entirely on my own, in the tropical fever forests, and another eighteen months in a shepherd's cot in the Welsh mountains, well within sight of Wordsworth's hills. Nature was equally severe and clement in both places. What would have disturbed me in either case, and maybe sent me off my head, would have been to imagine that nature outside was evilly or kindly disposed to me, and consciously stalking me with misfortunes or trying to bestow benefits on me; for those are the very superstitions that make savages so miserably unhappy; and they return in the form of fetishistic credulity, where Europeans have lost their faith, as in Huxley's case. The difference in the Latin-Mediterranean type of 'nature poetry', seen at its best in the *Georgics* of Virgil, is that there is no brown study, no Buddhistified blurring and blending with 'the Whole', to the detriment of all outlines and sane values. Where a Latin poet creates a darkness, as in Saint John of the Cross's *Dark Night of the Soul*, it is not for the sake of merging with that darkness alone, nor the losing of all contours in a brown study; it is simply to rid the mind of a less intense form of reality, so as to give it all the more power to seize the more intense reality of God. The proof is in the fact that this kind of mystical poetry makes you wide awake, and the other soothes you almost to sleep; and this is the difference between the mystic and the mistagogue. The nature poets of the Latin races tend to differentiate, to particularise, and even to anthropomorphise the objects about which they write, as Virgil did with his bees, bringing them out clearly from their background. The tendency of the northern poets is to let nature envelop them around with moss, clouds, weeds, and flowers, until we and they disappear in a dream of mental vegetation and vapour.

Lorca, who had known intense suffering in childhood and throughout the rest of his life, since he was never to know the normal command of his muscles and limbs, grew up amongst those countrymen who, different from the poetical excursionists, know the particular in nature, and not a single entity with a capital N. The things that attract his attention most as a poet are always in his immediate surroundings, with their peasants and their animal population of bees, butterflies, nightingales, cicadas, frogs, and lizards. The poems in his first collection were written between the ages of eighteen and twenty-two. In dealing with the smaller creatures of the earth, he affects a sort of Lilliputian minuteness, almost Franciscan in its intimacy, which in its detail reminds one of the exquisite treatment of bees by Virgil and Góngora, or the even more perfect treatment of the fable of the Town and Country Mouse by Horace in his satires. We are reminded of the pre-Romantic poets in our own literature, the exquisite Drayton of the *Nymphidia*, the speech of Mercutio about Queen Mab, and certain passages of *A Midsummer Night's Dream*. The very important difference is that we are torn by Lorca between a comic grotesqueness and a heartrending pathos with which he invests these slightly humanised, tiny creatures of the fields. He notices on the lizards their 'little white aprons', calls them 'drops of crocodile', and 'dragons of the frogs', seeing their 'green frock-coat of a devil's abbot'. But Lorca's small animal and insect world, though conceived with a childish directness of vision, is no dream world of Titanias and fairies, but the real old world we inhabit ourselves, seen in miniature, with startling clearness, as through the wrong end of a telescope. Its fierceness remains undiminished by microscopic proportions, and becomes all the more startling because of them and because of a sort of Goyaesque and Bosch-like mixture of the human with the Lilliputian.

The description of the voices of the frogs 'freckling the silence with little green dots' is another uncanny but perfect image to those who have heard the frogs of the southern marshes, at sundown, start up their chorus with the twinkling of the stars. When I say Lorca partly humanises his creatures, I do not mean that he detracts from their peculiar frogginess, lizardliness, or whatever it may be. The touch of humanity seems to enhance and emphasize their innate quality as frogs, snails or ants, by the sheer force of contrast. I have mentioned how Lorca used to love harrowing and terrifying the peasants by imitating the priest's 'hell-fire sermons'. Some of his poems, of a seemingly trivial nature, are so poignant that one sometimes indignantly and resentfully accuses the poet of going out of his way to make one suffer.

In one of his earliest poems describing the Odyssey of a snail, we are made to enter the nightmare world of suffering. The snail, 'the peaceful bourgeois of the meadow', feels a sudden curiosity about the world and decides on seeing, if he can, 'What is at the end of the path'. He meets two frogs, one of which is blind and both of which are beggars; after a depressing argument with them which brings on a mood of pessimistic doubt, he meets some ants. I quote this passage in a literal prose translation, since it is too hard to translate into verse:

> Now over the path
> An undulating silence
> Flows from the olive grove.
> With a group of red ants
> He next encounters.
> They are going along angrily,
> And dragging behind them
> Another ant with his
> Antennae clipped off.
> The snail exclaims:
> 'Little ants, have patience.
> Why do you thus illtreat
> Your companion?
> Tell me what he has done,
> And I will judge in good faith;
> Relate it, little ant.'
> The ant, by now half-dead,
> Says very sadly,
> 'I have seen the stars.'
> 'What are stars?' say
> The other ants uneasily.
> And the snail asks
> Pensively, 'The stars?'
> The ant repeats,
> 'I have seen the stars.
> I went up to the highest tree
> In the whole poplar grove,
> And saw thousands of eyes
> In my own darkness.'
> The snail asks again,
> 'But what are stars?'
> 'They are lights which we carry
> On the top of our heads.'
> 'We do not see them',
> The other ants remark.
> And the snail says 'My eyesight
> Only reaches to the grass.'
> The ants exclaim,
> Waving their antennae,
> 'We shall kill you

You are lazy and perverse;
To labour is your law.'
'I have seen the stars',
Says the wounded ant.
And the snail passes judgement:
'Let him go free,
Continue your work.
It's likely that soon,
Worn out, he will perish.'

Across the mild wind
A bee has passed.
The agonising ant
Inhales the vast evening
And says, 'It is she who comes
To take me to a star.'

The other ants run off
On seeing he has died.

The snail sighs
And goes off amazed
And full of confusion
At the eternal. 'The path
Has no end', he exclaims ...

This disconsolate mood recurs with insistence and power in Lorca's early work: and it returns even more powerfully in his later plays in which he invariably deals with what we call 'unpleasant subjects'. But in these later plays, the cruelty which he flings in one's face almost aggressively, as if to relieve his own sufferings, is tempered and balanced by the Euripidean stature of the protagonists, their innate strength, their willingness to accept suffering (since there is always a way out by cowardice or compromise) and their capacity for resignation. We accept his plays as we would accept them from scarcely anyone else dealing with the same subjects. It is his sheer artistic mastery which makes us accept them, and the realisation that his compassion is, after all, one of the main motives for burning both himself and us with such anguish.

In his *Romancero Gitano* we get the same cruelty and suffering. Benjamín Janés calls these poems 'Little pictures drawn in ice by a refined savage'. But in these perfect little pictures we are removed from the direct anguish of the emotional stab, since

Lorca is not telling us the tale himself, but, as it were, through the mouth of a gipsy, whom he is parodying and burlesquing at the same time. This removes the impact so that we receive it at second hand, as we do the very real tragedy of *Tam O' Shanter* in Burns, which if it had been written straight out in the vernacular by Tchekov, Maupassant, or Liam O'Flaherty would be an extremely sordid and harrowing tale. But Burns, whose native language was English, not Lallands (as we see from his letters to his intimates) puts the whole story into 'patois', and tells it through the mouth of a slightly naïf, comical personage, so that we receive the gruesomeness on a shield or cushion of laughter. In the same way the personality of the gipsy whom Lorca impersonates in all his native mannerisms and naïveté protects the reader from the full impact of the terrifying apparitions, the crimes, the brutal murders, the martyrdoms described in the *Romancero*. In the end it is clear that Lorca is not deliberately inflicting pain on the reader, in order to shock or annoy him; but that he feels so poignantly that he has to share this feeling with others. This is the motive underlying his insistence on themes of cruelty. We know that in his life he was cheerful, full of fun, a radiant and kindly personality, and that considering the extent and nature of what he had to suffer, there was not much perversity in his make-up, as modern poets go.

Together with this sense of pain one feels almost everywhere in Lorca's poetry, even at its gayest (and it can be very gay) there goes also the sense of a lurking, imminent and violent death. Even in his most vernal poetry, the shape and presence of death is always there, as in this *Spring Song*:

> On the lonely mountain
> A village cemetery
> Appears like a field
> Sown with seeds of skulls
> And cypresses have flowered
> Like gigantic heads,
> Which, with empty eye-holes
> And green hair,
> Pensively and sadly
> Contemplate the skyline.
>
> Divine April, who comest
> Charged with sunlight and perfumes,
> O fill with golden nests
> These flowering skulls.

Like Webster in Eliot's poem, Lorca is 'much obsessed by Death', and not only 'sees' but actually 'feels' and 'tastes' 'the bones beneath the skin'. In the *Ballad of the Little Square* (*Balada de la Placeta*) the children ask him:

> What do you taste in your mouth
> So ruddy and thirsty?

and he replies:

> The taste of the bones
> Of my enormous skull.

Again, in the most voluptuous of all Lorca's early poems, the *Canción Oriental*, which owes perhaps a debt to Paul Valéry's *Douces grenades entrouvertes* in which the jewelled seeds of a pomegranate are compared to thoughts ripening in the brow of a poet, we get a triumphant image of the pomegranate as half a heart and half a skull. This is one of Lorca's richest poems, glittering with the plunder of other poets such as Góngora, Valéry, and Darío, yet at the same time burning with Lorca's own personality, for he has already fully digested these outside influences; and the plunder already belongs to him by *right of conquest:*

> The fragant pomegranate! in it
> A heaven seems to crystallise
> (In every seed a star is lit
> In each red film a sunset dies).
> It seems a tiny hive that drips
> With live blood soaking through its mesh
> Because the bees have formed its pips
> Of women's mouths and kisses fresh:
> And when it bursts, a thousand lips
> Are laughing in its crimson flesh . . .
> The pomegranate is like the hoard
> Of the old goblin of the glade
> That in the pathless woods abroad
> Met with the solitary maid.
> It is the treasure whose red rays
> The green leaves guard within their hold,
> It is the ark of gems that blaze
> Within the dim-seen casque of gold.

The corn-ear is the bread. The Christ
In death or life lies there concealed.

The olive stands for hardness, spliced
With strength and labour in the field.

The apple is a carnal thing,
The sphinx's fruit, the food of sin,
The drop of juice that aeons wring
With Satan's touch upon the skin.

The orange burns with grief untold,
Grief that white blossoms were profaned,
Since now it's flushed with fire and gold
That was so spotless and unstained . . .

Chestnuts for peace by the fireside
And bygone things of yesterday:
The crackle of old logs, the sigh
Of pilgrims who have lost their way.

The acorn is the poetry
Of what is ancient and mature.
In the pale yellow quince we see
The cleanliness of health that's pure.

But in the pomegranate the fierce
Blood of the sacred heaven gleams,
Blood of the earth which waters pierce
With the sharp needles of their streams,
Blood of the boisterous winds that sweep
From the rough mountains that they rake,
Blood of the ocean's windless sleep,
Blood of the hushed and drowsy lake.
The pomegranate is the prehistory
Of our own blood. So gashed apart,
Its bitter globe reveals the mystery
Both of a skull and of a heart . . .

The obsessional presentiment of death is pre-eminent even in Lorca's most lush and pastoral descriptions. It is to be remembered that death was very much in the air in the twenty years that preceded the Civil War. 'Viva la muerte', the cry of the anarchists, was frequently heard, and their skull and crossbones chalked up everywhere. The sense of death is deeper in Lorca than in most other Spanish contemporary artists. Someone has said jestingly that Death is the patron saint of Spain; and Barrès says that the Spanish consciousness is founded in voluptuousness, blood, and death. To accept this would be to discount such seraphic personalities as Saint John of the Cross, Saint Teresa of Avila, Saint Francis Javier, and Saint Ignatius of Loyola, who are among the most ethereal and purely spiritual creatures in history. The great Castilian mystics had transcended death. Their attitude to death is the very opposite of Lorca's: it is one of longing for death not out of tiredness but because of an intenser life promised after death.

> I die because I do not die

says Saint Teresa, expressing the very opposite of the terrified thrill of anguish which Lorca expresses at the thought of death, and which Goya depicts with such violent force and horror.

In Lorca's first book of verse, *Libro de Poemas*, is a most interesting experiment in the shape of an *Elegy to Lady Joan the Mad*. Joan the Mad was the daughter of the great queen Isabella; she was driven mad partly by lover's jealousy for her warrior-husband; and in this elegy, while suppressing any historical allusions which would tend to anachronise or 'date' the theme, Lorca tackles the subject in a grandiose manner, with full orchestration, almost as if he were toying with the idea of the epic treatment of some such subject. It is powerfully moving and reminds one for a moment of Camoes's passage referring to the unfortunate historical figure of Inez de Castro in the *Lusiads*; or of the broad full stream of Espronceda's rhetoric in his *Canto a Teresa*. The following lines are selected from the closing verses of the elegy:

> You had that passion which the sky of Spain confers.
> The passion of the dagger, the listening ear, and the dirge.
> O divine princess of the crimson twilight
> Whose spinning-wheel was iron, whose thread was steel.
>
> You never knew the bower or the sad madrigal,
> Nor the troubadour's praises sobbing in the distance.
> Your troubadour was a lad with silver scales
> And the echo of a trumpet was his wooing.
>
> Yet without doubt you had been formed for love,
> Made for the sigh, for spoiling, and for fainting,

To weep your grief on a beloved breast
While you tore a scented rose between your lips . . .

Granada keeps you like a holy relic,
O dusky princess sleeping in the marble.
Heloise and Juliet were but as two daisies,
But you a red carnation full of blood
Who came from the golden earth of Castile
To sleep between the snows and the chaste cypresses.

Granada was your deathbed, Lady Joan,
She of the ancient towers and the still-hushed garden,
She of the ivy dead on the red walls,
She of the blue mist and the romantic myrtles.

Princess that loved without the due reward.
Red carnation in a deep and desolate valley.
The tomb that keeps you oozes forth your sorrow
Through eyes which it has opened in the marble.

These powerful alexandrines remind one of the vitality and verve of Rubén Darío. In spite of some immaturities such as the use of 'poetical' words, like 'romantic myrtles', there is a surge of movement and colour in this poem which makes one wonder why Lorca never returned to this type of elegy orchestrated with epic undertones, like Shelley's *West Wind*. But it seems that already in his first book of verse Lorca had defined for himself the limits within which he was to write from then onwards; and the only time he ever approaches the same broad and solemn treatment of such a theme is much later in the *Three Odes*, two to the Blessed Sacrament, and one to Salvador Dali, though these are sustained by a succession of unconnected images which do not fuse into one another dynamically, as here in this elegy with its cumulative and one-way stream of images, increasing in momentum.

Everywhere in the *First Poems* one is conscious of Lorca's growing powers of imagery. Sometimes the imagery is there for its own sake; and sometimes there is a slightly obsessional repetition as when the 'chopo' or black poplar tree, with its swaying gestures, is more than once compared to an old man, a music-master, or a schoolmaster, and once we see it aiming a box at the ear of the moon for disturbing a music party of frogs, crickets, and trees. One wonders of what teacher or master it is the affectionate souvenir. Poplar trees are so much a part of the landscape in Andalusia, with their perpetual nervous fidgeting, that instead of getting annoyed by the repetition (for Lorca seldom makes a repetition inadvertently) we begin to feel an amused affection for the presiding schoolmaster of so many of his evening classes, group parties, and concerts of trees. Here is a brief passage in which a mere visual fancy or image sustains and performs the function of a poem, like one of those short Chinese or Japanese poems which are satisfied with producing a single vivid picture:

> Trees.
> Have you been arrows
> Let fall from the azure?
> What terrible warriors shot you forth?
> Were they the stars?

He addresses the passion-flower as the 'anvil of the butterflies' and mosquitoes as the 'Pegasi of the dew'. A straight road is a 'lance wounding the horizon'. Donkeys, the most fatalistic and resigned of creatures, are called 'Buddhas of the fauna', and the high road with its countless tracks, and spoors, and cart ruts, an 'enormous cheiromancer' and the 'Flammarian of foot-marks'. Throughout the later poems in the *Libro de Poemas* we feel Lorca perfecting his command of these terse and vivid epithets, of which he showed such mastery in his later work that they become his very style itself.

In poems such as *Prólogo* Lorca affects a faintly diabolical and Byronic attitude, probably through the influence of Baudelaire, Lautréamont, and perhaps Salvador Dali in his early youth; but in this swashbuckling pose, as in most of his erotic poetry, until we come to that one brilliant exception, the *Casada Infiel*, and to love-scenes in his plays which treat love objectively, Lorca is self-conscious and ill-at-ease. He gave up this occasional Byronic pose after his first book; why it did not suit him is probably explained in his self-analysis in one of his very earliest poems where we see him over-sensitive and gentle, and something like *El Desdichado* in Gérard de Nerval's famous sonnet:

> The waif, the shade, whose grief is absolute,
> The prince of Aquitaine whose tower fell down.

Now follows Lorca's portrait of himself. It is instinct with a deep sorrow which his outwardly radiant and pucklike cheerfulness concealed from his friends:

> I go weeping down the street
> Grotesque and bewildered
> With the sorrow of Cyrano
> And of Quixote,
> The redeemer
> Of impossible infinities.
> With the rhythm of a clock
> I watch the lilies wither
> At the contact of my voice,
> And in my lyrical song
> I wear the trappings
> Of dusty clown. Love,
> So beautiful and handsome,
> Has hidden under a spider. The sun
> Like another spider hides me
> In tentacles of gold. No!
> I'll never prosper in my venture
> Because I am like Love himself
> Whose shafts are lamentations
> And his quiver the heart.

There is no dramatised self-pity in this portrait. Both Lorca's suffering and his capacity for suffering were very great. He did not suffer morbidly, but as a true poet should, by trying to turn his suffering into poetry, and he did this better than most other contemporary poets. His eternal wrestling with the theme of death had its justification, just as Keats's preoccupation with the same theme. It came out of strength rather than weakness. The event proved that it was no illusion. His own violent death, and that of nearly three million Spanish men, women, and children, could already be sensed in the air, like a coming thunderstorm, for many years before the Terror was unleashed. It was publicly threatened on all the walls. One could not escape from being confronted with skulls and crossbones with bloodthirsty inscriptions chalked up everywhere. In his *Song for the Moon*, (*Canción para la Luna*) and in many other poems, he senses and prophetically describes the event, playfully apostrophising the moon, and yet clearly foreseeing the desolation of his country. Even in this playful piece of moonshine, we feel the imminence of disaster:

> Living lesson
> For anarchists!
> Jehova has the habit
> Of scattering his farmyard

With dead eyes
And the little heads
Of contrary
Militias . . .
Live in the hope,
Dead eyeball,
That the great Lenin
Will be the Big Bear
Of your landscape,
The bleak ridge
Of the sky
Which will tranquilly drift
To give the last embrace
To the Old Man
Of the Seven Days.

And then, O moon,
So white, will come
The unsullied reign
Of dust and ashes.

In this and other passages is a definite fore-sense of the useless chaos about to be precipitated over Spain. In the repeated wrestling with the idea of death, Lorca generally increases the stature of life, and intensifies it. All those people who repeatedly seek out death to risk their lives do so chiefly because they are overflowing with a surplus of life. They get a stimulus from the presence of death as a healthy body does from a cold bath.

In Lorca's time, the men who risked their lives most consistently were the priests and the great matadors; several of the latter were killed in the exercise of their profession; as were the majority of the monks and priests also, who did not even receive the pay of the lowest artisans for the most dangerous vocation of all. In Lorca's greatest sustained lyrical poem, the *Lament for the Death of the Matador* (*Llanto por Ignacio Sánchez Mejías*), we see a duel between Life and Death, enacted almost as a ritual dance between the superb, overflowing vitality of the matador and the cold shadow of Absence, while each augments and enhances the stature and the mystery of the other. In all his early poems in which he treats the subject of death we feel that the dual process is at work, which he carries to such a supreme triumph in the *Llanto*. Many of the earlier poems are rehearsals of this towering spiral in which the two forces contend in a sort of ecstasy; but there are also heights of serene lyrical contemplation which are exceptional in such a young poet, above all in these lines from the poem *Mañana* (*Morning*):

> But the song of water
> Is an eternal thing.
>
> It is light become the sound
> Of romantic illusions.
> It is firm, yet soft,
> Meek, and full of heaven.
> It is the mist and the rose
> Of the eternal morrow.
> Honey of the moon which flows
> From buried stars.
> For some good reason Jesus
> Realised himself in water.
> For some good reason Venus
> In its breast was engendered.
>
> Christ must have said to us:
> 'To whom better, my brothers,
> Can we confide our sorrows
> Than to her who rises up to heaven,
> Arrayed in a spiral of whiteness?'

Ideas are continually suggested to Lorca by the sound of water, to whose endless dropping he compares the sound of a far-off guitar in one of his finest poems, the *Cante Jondo*, which is quoted in full on page 70. The sound of rain in the poem *Lluvia* evokes the following significant passage concerning the fall of rain:

> It is the dawn of fruit. It is that which brings us flowers
> And anoints us with the holy spirit of the seas,
> That which sheds life over the down lands
> And in the soul a sorrow which is not known.
>
> The terrible nostalgia for a lost life
> And the fatal sentiment of having been born too late,
> Restless illusion of an impossible tomorrow
> With the close inquietude of fleshly pain.
>
> Love is awakened in the greyness of its rhythm.
> Our interior sky contains a triumph of blood.
> But all our optimism turns to sorrow

To contemplate the dead drops on the glass.

And those drops are eyes of the infinite; gazing
Back into the white infinity which is their parent.

Each drop of water trembles on the dim glass
Leaving divine wounds of diamond.
They are the poets of water who have seen and meditate
Things which the vast crowds of rivers ignore.

That image of the inactive drop of water filled with conscious light as contrasted with the blind strength of the brawling rivers gives us an idea of Lorca's conception of the poet's function in the scheme of things—that of static inward illumination, lit up even by one's own sorrow.

III

The 'Romancero Gitano'

'Romancero' means a collection of 'Romances'. This is by far the most famous and popular of all Lorca's collections of verse. Lope de Vega, greatest of all Spanish poets and dramatists in the sixteenth century, said 'las relaciones piden romances', 'narratives require the romance form'. The romance is the Spanish equivalent of our ancient popular ballad-form; and the latter is the only means by which we can possibly translate it, since the Spanish 'romance', instead of rhyming, requires an assonance in every second line, which continues right through the whole poem, and, in the long run, produces a stronger effect than our rhyme. Although an ancient form, it is still most serviceable for modern narration. It would be almost impossible to imitate in English, since, even where we could produce a sequence containing the necessary vowels, there is too much variation in the manner in which we pronounce them to get anywhere near the effect produced by the Spanish assonance. For instance, let us take the double assonance of the letters 'i' and 'e' in the word 'winter'. We could assonate the word 'filbert' with it: but the words 'bridges' and 'midget' would be disqualified, since the sound of the 'e' in 'winter' is like that of the letter 'u' in 'cup', and the sound of the 'e' in 'bridges' is like the 'i' in 'is'. Then we come to the words 'silence' and 'trident' with their 'i' and 'e', and we should be completely at sea. We have not enough similar word-endings in our language for 'romance' to be practical. The best we can do in English is to serve up an attempted equivalent in our own language, and the nearest we can get to the romance form is a loose tetrameter quatrain, rhyming in the third and fourth lines; this reproduces the rhythm of the original in marked periods of two lines.

Lorca, who while we are reading him appears to be a very spontaneous and facile writer, must have expended prodigious study and pains on the Romance form; his assonance is very much more pronounced than that of other Spanish poets. On analysing his use of vowel sounds, one finds that the extra emphasis which his own vowel assonance sounds acquire is usually due to their having been entirely suppressed from the rest of the line in which they occur. This is a true feat of profound technical engineering. The things one notices least are often the things that matter most in poetry. The poetry which is the hardest to write is almost invariably the 'easiest' to read, and vice versa. Very few poems in modern verse are so immediately and pleasantly readable to all classes of readers as are those of *Romancero Gitano*, for although they often deal with tough and unpleasant subjects, burlesque and laughter act both as an antiseptic and an anaesthetic. The first of these poems, *The Ballad of the Moon*, is an almost childish fancy. In a way it reminds us of Blake's *Little Boy Lost*. The dream of the slumbering child, in the end, becomes the reality of the poem:

> The moon came to the farrier's shop
> Wearing her bustle sprigged with nard.
> The little boy is staring at her,
> The little boy is staring hard.
> The moon is waving her white arms
> Into the palpitating air,
> And shows, lascivious yet pure,
> Her breasts of tin so hard and bare.
> 'Escape from here, O moon, the moon,
> For if the gipsies come in sight
> They'll take your heart and make of it
> Necklets of beads and trinkets white.'
> 'Child, let me be, leave me to dance,
> For when the gipsies come at last
> They'll find you sleeping on the anvil
> With your little eyes shut fast.'
> 'Escape from here, O moon, the moon,
> I hear their horses in the night.'

> 'Leave me, child, and do not trample
> My whiteness with its starch of light.'
> Approaching fast a horseman beat
> His drum, the plain, with rolling tread.
> The child was lying on the anvil
> With eyes shut fast as she had said.
>
> Along the olive orchard came,
> All bronze and dream, the gipsy set,
> With heads uplifted proudly high,
> And eyes half-closed, like slots of jet.
>
> O, how the night-jar sang that evening
> Up in the tree-tops loud and high,
> While hand in hand the moon is leading
> The little child across the sky.

In the last verse the gipsies come to the forge, and start weeping and giving loud cries on not finding the child, but the last verses seem to lose their force since our attention is transposed from the moon to the wind, which in its turn is supposed to be protecting and keeping guard over the moon.

The second ballad, *Preciosa y el Aire* (*Preciosa and the Wind*), has been claimed as Freudian territory by the psychoanalytic school of poets; but Freud, after all, did not invent the sexes, nor nervous hallucinations on the part of young girls; and it would not be the first time that a girl, getting a fright in the dark, has rushed off in a panic to the nearest company she can find. This ballad, beginning in a dream atmosphere, with the fright of the gipsy girl, ends the opposite way round from the last (which ends in a dream), namely in the matter-of-fact atmosphere of the British Consul's house. The figure of Saint Christopher is used in Spanish to describe any extremely hirsute, muscular, vigorous, and swarthy type of person:

> Beating upon the moon of parchment
> Preciosa with her tambourine
> Comes down by an amphibious path
> Of laurel shade and crystal sheen.
> The silence bare of any star,
> Scared by the jangled sound she rings,
> Falls where the deep sound of the ocean
> Starry with fish, resounds and sings.
> Amongst the peaks of the sierra

Slumber the coast-guard carbineers
Keeping a watch upon the towers
Where English folk have lived for years.
Beating on her moon of parchment,
Preciosa comes with rhythmic fall;
To see her come the rude wind rises,
The wind that does not sleep at all.
A huge Saint Christopher stark naked
Full of celestial tongues of air,
He looks upon the girl, and plays
On a sweet pipe that isn't there.
'Allow me, girl, to lift your skirt
And let me see you plain and clear.
Open to my ancient fingers
The blue rose of your beauty, dear!'

Preciosa flings away her tambour,
And runs, and runs, and does not tire
And the Big-Man-Wind pursues her
With a burning sword of fire.

The sea has puckered up its rumour,
All pale as death the olives grow.
The shrill flutes of the shadows sing.
So does the smooth gong of the snow.

Preciosa run! or the green wind
Will surely have you by the hair!
Run, Preciosa! run like mad!
Look out! He nearly got you there!
The satyr of the setting stars
With all his glittering tongues of air.

Preciosa, terrified to death,
Runs into the first house she sees,
Where high above the lofty pines,
The English Consul lives at ease.

Alarmed to hear her piercing screams
Come rushing down three carbineers
With their black cloaks hugged tightly round them
And caps pulled down about their ears.

A tumbler full of lukewarm milk

> The Englishman provides in haste
> And a goblet full of gin
> Which Preciosa will not taste.
>
> And while she tells her story weeping
> And they are listening, without pause
> Against the roof-top tiles above them
> The wind in fury gnashed his jaws.

Preciosa is the name of the beautiful gipsy singer and dancer in Cervantes's *Novelas Ejemplares*, and its use here is probably a conscious debt to the Master.

In *Reyerta*, we leave the region of dreams and visions for a matter-of-fact brawl between gipsies. It is related naïvely, though intended to be read ironically, as where the Judge is quoted as saying 'Five of the Carthaginians have been killed, and four Romans', meaning of course that it was a very ancient feud (and one of which he, as Judge, was heartily sick) between two clans of gipsies. In the Catholic way of thinking prevalent among peasants, those who minister to the injured and wounded are, for the moment, performing the ministry of angels; in this poem, and also later in the *Romance of the Death of Antonito el Camborio*, those who are succouring the distressed need not necessarily refer to supernatural angels, but gipsy women, and their 'wings' would be the long shadows of their shawls on the late afternoon hillside. The 'knives of Albacete' refer to long clasp-knives which are manufactured cheaply and excellently in that town, and sold throughout Spain; their blades are often set with a ratchet to stop them from folding up while fighting, which gives the comparison to angels' wings here a slightly more humorous twist. The blades are very long and gracefully curved, like a Cupid's bow, at the end. As wings they would, in shape, resemble those of shearwaters, and of many angels seen in mediaeval sculptures, whose wing-joints when folded rise higher than their heads, and the tips of the wings taper very sharply to their heels. The reference to the bull of altercation 'climbing the walls' is sheer fun. 'To climb the walls' in Spanish means to 'go off the deep end'; this is an intentional malapropism. The decorative allusion to knife-blades as lilies ought not to be missed.

> In the midst of the ravine,
> Glinting Albacete blades,
> Beautified with rival bloods
> Flash like fishes in the shades.

A hard flat light of playing cards
Outlines, against the bitter green,
Shapes of infuriated horses
And profiles of equestrian mien.
Under the branches of an olive,
Weep two women bent with age,
While the bull of altercation
Clambers up the walls with rage.
Black angels come with handkerchiefs
And water from the snowline-boulders,
Angels with vast wings, like the blades
Of Albacete, on their shoulders.
Juan Antonio from Montilla
Down the slope goes rolling dead,
With his flesh stuck full of lilies,
A sliced pomegranate for his head;
And now the cross of fire ascends
Along the highways of the dead.

The Judge and Civil Guard their way
Along the olive orchard take,
Where slithered blood begins to moan
The dumb song of an injured snake.
'Gentlemen of the Civil Guard!
The same old story as before—
Five of the Carthaginians slain
And of the Roman people four.'

The maddening afternoon of figtrees
And of hot rumours, ending soon,
Fell down between the wounded thighs
Of the wild horsemen in a swoon.
Black angels fly across the air
From which the setting sun departs,
Angels with long dark streaming hair
And oil of olives in their hearts.

From the most ancient times down to the modern, the repetition of the word 'verde', green, has haunted the Spaniards in various refrains from the ancient ballad:

> Rio verde, rio verde
> Mas negro vas que la tinta!

Bécquer, and the doyen of living Spanish modern poets, Juan Ramón Jiménez, who has had a considerable effect on Lorca, have both played magically with the repetition of the word 'verde' in a similar manner. In the *Romance Somnámbulo* (*Somnambulistic Ballad*) Lorca uses it to produce a ghostly atmosphere of moonlight in which the vague encounter of the two gipsies occurs. The story is embryonic, and hardly exists at all outside the conversation of the dying man and the father of the gipsy girl. In fact, one is not sure whether the two male figures are not already spectres—but it hardly matters. This is one of the most famous of Lorca's romances:

> Green, green, how deeply green![1]
> Green the wind and green the bough,
> The ship upon the ocean seen,
> The horse upon the mountain's brow.
> With the shadows round her waist
> Upon her balcony she dreams.
> Green her flesh and green her tresses,
> In her eyes chill silver gleams.
> Green, green, how deeply green,
> While the gipsy moonbeam plays
> Things at her are gazing keenly
> But she cannot meet their gaze.
>
> Green, green, how deeply green!
> See the great stars of the frost
> Come rustling with the fish of shadow
> To find the way the dawn has lost.
> The fig-tree chafes the passing wind
> With the sandpaper of its leaves,
> And hissing like a thievish cat,
> With bristled fur, the mountain heaves.
> But who will come? And by what path?
> On her verandah lingers she,
> Green her flesh and green her hair,
> Dreaming of the bitter sea.
>
> 'Companion, I should like to trade
> My pony for your house and grange,
> To swap my saddle for your mirror,

---

[1] Literally 'Green, green, I want you green': but it has this secondary meaning too.

My sheath-knife for your rug to change.'
'Companion, I have galloped bleeding
From Cabra's passes down the range.'
'If it could be arranged, my lad,
I'd clinch the bargain; but you see
Now I am no longer I,
Nor does my house belong to me.'
'Companion, I should like to die
Respectably at home in bed,
A bed of steel if possible,
With sheets of linen smoothly spread.
Can you not see this gash I carry
From rib to throat, from chin to chest?'
'Three hundred roses darkly red
Spatter the white front of your vest.
Your blood comes oozing out to spread,
Around your sash, its ghostly smell.
But now I am no longer I
Nor is my house my own to sell.'
'Let me go up tonight at least,
And climb the dim verandah's height.
Let me go up! O let me climb
To the verandah green with light!
O chill verandahs of the moon
Whence fall the waters of the night!'

And now the two companions climb
Up where the high verandah sheers,
Leaving a little track of blood
Leaving a little trail of tears.
Trembling along the roofs, a thousand
Sparkles of tin reflect the ray.
A thousand tambourines of glass
Wounded the dawning of the day.

Green, green, how deeply green!
Green the wind and green the bough.
The two companions clambered up
And a long wind began to sough
Which left upon the mouth a savour
Of gall and mint and basil-flowers.
'Companion! Tell me. Where is she?
Where is that bitter girl of ours?'
'How many times she waited for you!

> How long she waited, hoped, and sighed,
> Fresh her face, and black her tresses,
> Upon this green verandah-side!'
>
> Over the surface of the pond
> The body of the gipsy sways.
> Green her flesh, and green her tresses,
> Her eyes a frosty silver glaze.
> An icicle hung from the moon
> Suspends her from the water there.
> The night became as intimate
> As if it were the village square.
> The drunkards of the Civil Guard
> Banging the door, began to swear.
> Green, o green, how deeply green!
> Green the wind and green the bough,
> The ship upon the waters seen,
> The horse upon the mountain's brow.

The *Gipsy Nun* (*Monja Gitana*) is a very pleasant piece of verbal embroidery. We notice here a similar image to that in which Saint Gabriel says to the Virgin 'Your eyes are like the burning plains wherethrough the lonely horsemen wend' in the poem about Sevilla. Here Lorca writes:

> Within the dark eyes of the Nun
> Two horsemen gallop . . .

The final quatrains referring to the perpendicular embroidery on the horizontal landscape of her sampler are charming, and so is the vision of the sunlight playing chess with the leaves outside her lattice:

> Under the blaze of twenty suns
> How steep a 'level plain' inclines!
> What rivers running vertical
> Her burning fantasy designs!
>
> But she continues with her flowers,
> While, standing upright in the breeze,
> The sunlight plays a game of chess
> Over her lattice with the trees.

The *Romance de la Casada Infiel* is the best known of all

Lorca's poems because of its erotic appeal. But this should not be allowed to eclipse some of the finest descriptive passages as the lovers go down the river bed:

> It was on Santiago's Eve
> As by an obligation made;
> The stars had all gone out; the crickets
> Lit up and twinkled in the shade.
>
> The trees grew taller at their tops
> Wherein no light was seen to quiver
> And the horizon, loud with dogs,
> Was barking far across the river . . .
>
> Not spikenard nor conch of pears
> Can boast a skin so fine and clean,
> Nor can the glass of moonlit mirrors
> Reflect so crystalline a sheen.
> Her thighs escaped me in the dark
> Like startled fishes silver-shoaled,
> Half of them shimmering with fire,
> Half of them shivering with cold.
> Upon that night I journeyed there
> The finest road of all earth
> Galloping on a mare of pearl
> Without a stirrup or a girth!

The gipsy who recounts this adventure comes right down with a bathotic lapse at the end (which is quite intentional on Lorca's part) when he presents the 'unfaithful wife', whom he had suspected of being unmarried, with a workbasket, and decides that as she is a married woman, he must be careful not to fall in love with her.

In the *Romance de la Pena Negra* (*Ballad of the Black Sorrow*), Lorca attempts to personify, in the shape of Soledad Montoya, a gipsy, whom he addresses, the immemorial sorrow of the gipsy people:

> O Soledad of all my sorrows,
> Like a stampeding horse that raves
> And when it meets the sea at last
> Is swallowed outright by the waves!
> 'Do not remind me of the sea
> That with the same black sorrow grieves

> Over the country of the olives
> Under the rumour of the leaves.'

> In the fresh water of the larks
> Refresh your body, and release
> Your weary heart, O Soledad
> Montoya! to repose in peace.

> Away down there the river sings
> The skirt-flounce of the sky and leaves.
> Crowning itself with pumpkin flowers
> The new light rustles through the sheaves.
> O sorrow of the gipsy people,
> Clean sorrow lonely as a star,
> O sorrow of the hidden fountain
> And of the daybreak seen afar!

The next three Romances, *San Miguel* (Granada), *San Rafael* (Córdoba) and *San Gabriel* (Sevilla), lovingly burlesque the somewhat highly coloured statues of the patron Saints of Granada, Córdoba and Sevilla, a trinity of cities which often recur side by side in single poems of Lorca's. Religion in the South expresses itself chiefly in colour, rhythm, and dances, and the style of the Church images in Andalusia is in full accordance with the flamboyant taste of the gipsies.

Saint Michael looks out of his alcove in the high tower of Granada Cathedral, into the sky of snowpeaks:

> From the verandahs they are seen
> Along the rocky mountain tracks—
> Mules, and the shadows of the mules,
> With loads of sunflowers on their backs.

> His eyes amongst the shadows
> Are tarnished with enormous night
> And up the spirals of the air
> Passes the dawn with salty light.

> A sky of mules as white as milk
> Closes its glazed, mercurial eyes
> Imposing on the twilight hush
> A period to hearts and sighs.
> The water makes itself so cold
> That nobody to touch it dares,

Mad water, running naked stark
Along the rocky mountain stairs.
Saint Michael, laden with his laces,
In the church-alcove where he camps
Is showing off his lordly thighs
Surrounded by a ring of lamps.

Archangel of domestic meekness,
When the stroke of midnight rings
He feigns a sweet fictitious anger
Of nightingales and rustling wings.
He sings amongst the stained glass windows,
Ephebus of three thousand eves
Fragrant with water of Cologne
But far away from flowers and leaves.

Waves on the shore compose a poem;
Each in its window-bay rejoices:
The river borders of the moon
Lose in reeds to gain in voices.
Flashy 'manolas' from the slums
Come chewing sunflower seeds and pips
With their occult, enormous bums
Like brazen planets in eclipse.
Tall gentlemen come down the way
With ladies sorrowful and frail
Wan with the thoughts of yesterday
And memories of the nightingale.

And the Bishop of Manila,
So poor and saffron-blinded, then
Says a Mass which has two edges
One for the women, one for men.

Saint Michael stayed content and quiet
Up in his garret in the tower
In his skirts, cascading finery,
Where crystals, lace and trinkets shower,
Saint Michael, ruler of the lamps,
And of the Offices and Paters,
Poised in the Berber eminence
Of crowds and wondering spectators.

The *Romance of Saint Rafael,* the patron Saint of Córdoba, is harder to translate, with its bathers and its basket-weavers by the riverside, beside other features which are not common to the landscape of the English speaking countries. The thing to remember is that the two Córdobas contrasted in this poem, between which a fish on the surface of the water acts as a bridge, are Córdoba, the real city, and Córdoba, the reflection in the water. The austere Romanity of the city of Seneca and Lucan is beautifully brought out:

1.

Along the riverside of reeds
Closed carriages assemble, where
The waves are polishing the bronze
Of Roman torsos brown and bare:
Carriages that the Guadalquivir
Portrays upon her ancient glass
Between the colour-plates of flowers
And thunders of the clouds that pass.
The lads are weaving as they sing
The disillusion of the world
Around the ancient carriages
By the encroaching darkness furled.
But Córdoba stirs not, nor trembles
Under the mystery they invoke,
Since, if the darkness were to shift
The architecture of the smoke,
With marble foot she reasserts
Her glory spotless and severe.
A flimsy petal-work of silver
Encrusts the breeze so grey and clear
Above the great triumphal arches
Displayed upon the atmosphere.
And while the bridge sighs out its ten
Reverberations of the sea,
Contrabanders of tobacco
Between the broken ramparts flee.

2.

A single fish within the water,
Links the two Córdobas and joins
The gentle Córdoba of reeds
To that of architraves and groins.
Lads with expressionless blank faces
Along the bank strip to the skin,

Apprentices of Saint Tobias
And belted rivals of Merlin,
To tease the fish with taunting queries
Whether it would prefer more soon
Red splashes of the flowers of wine
Or acrobatics of the moon.
But the fish that gilds the water
And makes the marble dark and solemn,
Instructs them in the equilibrium
Of a solitary column.
The Archangel, arabianised,
With gloomy spangles all around,
In the mass-meeting of the waves
Sought out a cradle in their sound.

A single fish within the water,
Two Córdobas in beauty clear.
Córdoba broken into streams.
Córdoba heavenly and austere.

In the *Romance of Saint Gabriel*, the patron Saint of Sevilla, Lorca gives a charming account of the visitation of our Lady by Saint Gabriel as it would be seen through the semi-pagan eyes of a wandering gipsy:

1.

A lad as graceful as a reed
With shoulders broad and body slight,
With a skin of moonlit apples,
Sad mouth, and large eyes brimmed with light,
Like a nerve of burning silver
Rounds the deserted street and square;
His shining shoes of patent leather
Trample the dahlias of the air
With their two rhythms that resound
Celestial dirges as they pace.
On all the seacoast is not found
A palm to equal him in grace,
Nor emperor that wears a crown,
Nor any wandering star in space.
When to his jasper breast he stoops
His forehead in that pensive way,
The night seeks out the lowliest plain
Because she wants to kneel and pray.

For the Archangel Gabriel
Lonely guitars sing on the breeze,
The tamer of the turtle-doves
And enemy of the willow-trees.
—Saint Gabriel, the child is weeping
Within his mother's womb alone.
Do not forget the suit of clothes
The gipsies gave you as your own.

2.

Annunciation of the Kings,
So richly mooned, so poorly dressed,
Opens the door into the street
To entertain her starry guest.
The archangel Saint Gabriel,
Between a lily and a smile,
Great-grandson of the high Giralda,
Had been approaching all this while.
In the embroidery of his jacket
The crickets palpitate and sing
And all the stars that lit the night,
Turning to bells, began to ring.
'Saint Gabriel, you see me here
Pierced with three nails of fierce delight.
Your glory from my burning face
Suns forth the jasmines opening white.'
'God is with you, Annunciation,
Brown beauty of the gipsy kind,
You'll have a son more beautiful
Than rushes waving in the wind.'
'Saint Gabriel, dearer than my eyes,
Dear Gabriel of my days and hours!
To seat you here I visualise
A bank of sweet carnation flowers.'
'God is with you, Annunciation,
So richly mooned, so poorly dressed,
Your son will have a little mole
And three red gashes on his chest.'
'Saint Gabriel, how your glory shines!
Dear Gabriel of my life and veins!
Down the bottom of my breasts
I feel the warm white milk that drains.'
'God is with you, Annunciation,
Mother of dynasties without end!

Your eyes burn like the barren plains
Through which the lonely horsemen wend.'

The baby sings within the breast
Annunciation to surprise.
Three seedlets of the almond green
Are trembling in his tiny cries.
Saint Gabriel through the silent air
Went up a ladder to the sky;
And all the stars of night were turned
To everlasting flowers on high.

For movement, verve, humour, and pathos, the next ballad, *Muerte de Antonito el Camborio*, is unequalled in Lorca:

Voices along the Guadalquivir,
Were heard. Old voices, croaking death,
Surround and trap the manly voice
With the carnations in its breath.
He bit the boots that stove his ribs
With slashes of a tusky boar.
He bucked the soapy somersaults
Of dolphins, slithering in his gore.
He dyed in his opponents' blood
The crimson necktie that he wore,
But then there were four knives to one
So in the end he could no more.
When in the grey bull of the water
Stars strike their javelins; in the hours
When yearling calves are softly dreaming
Veronicas[1] of gillyflowers,
Voices of death re-echoed screaming
Along that river bank of ours.

Antonio, of Camborio's clan,
That have blue manes both thick and strong,
With olive skins, like moonlight green,
And red carnations in their song,
Beside the Guadalquivir's shore,
Who took your life, who could it be?
'The four Heredias, my cousins,
The children of Benamejí.

---

[1] Veronica is a pass in bullfighting.

Things which they did not grudge to others
Were things for which they envied me—
My shoes of bright Corinthian hue,
My medals made of ivory,
And this fine skin, in which the olive
And jasmine both so well agree.'
'Alas, Antonio el Camborio,
So worthy of an empress high,
Remember now to pray the Virgin
Because you are about to die.'
'Ah! Federico García Lorca,
Go quickly while there's time, and raise
The Civil Guard for I am broken
And wilting like a stalk of maize.'

He had three leakages of blood
And then, in profile, there he died,
Live currency of gold whose like
Can never be again supplied.
A withered angel came and placed
A pillow underneath his head,
While others with a weary flush
Lit up a candle for the dead.
And when the four Heredia cousins
Back to Benamejí had come,
Voices of death along the river
Ceased to be heard: and all was dumb.

The next Romance is called *Muerto de Amor* (*He Died of Love*). This ballad is one of Lorca's excursions beyond the limits of ordinary experience into some sort of realm of his own imagining.

'What is that thing that blazes there
Along the corridors in heaven?'
'Come in, my lad, and close the door;
Already it has struck eleven.'
'Within my eyes, against my will,
Four blazing torches seem to pass.'
'It must be that the people yonder
Have started polishing their brass.'

A garlic-slice of sickly silver,
It seems the waning moon has thrown

Heads of hair with yellow tresses
Over the towers of yellow stone.
The night along the balconies
Calls trembling at the window-glasses,
Bayed after by a thousand dogs
Who do not know her as she passes.
A scent of ambergris and wine
Floats from the corridors on high.

Breezes from the dewy reeds,
With many a lost archaic cry,
Reverberated in the broken
Archway of the midnight sky.
Oxen and roses were asleep,
But through the corridors, four lights,
With all the fury of Saint George,
Vociferated in the heights.
Sad women from the valley came
Bearing their manly strength of blood,
Assuaged in the cut flower, and bitter
In the thighs of youthful bud;
Old women of the riverside
Wept in the valley for the flames
Of the intransitible moment
Of waving hair and whispered names.
Façades of whitewash cut the dark
And squared it off, abrupt and white.
The seraphim and gipsy people
Played their accordions in the night.

'Mother, when I am dead, to all
The gentlefolk proclaim it forth.
Let azure telegrams be sent
Travelling from the South to North.'
Seven cries, and seven bloods,
And seven poppies (double blooms)
Smashed the unreflecting mirrors
That tarnished in the darkened rooms.
Full of amputated hands
And funeral wreaths, in vast despair,
The ocean-flood of perjured oaths
Was thundering—I don't know where.
Heaven slammed its doors against the rumour
With which the forests heave and cry,

While the four lights vociferated
Along the corridors on high.

The following *Romance de la Guardia Civil Española* (*Romance of the Civil Guard of Spain*), is one of the longest in the book. The Civil Guard are actually a *corps d'élite* who lost about eighty per cent of their personnel during the Civil War. They wear tri-corned hats and blue capes, and are great fighters. They are detested by the gipsies as they uphold order on the country roads and are interested in keeping down the theft of horses and poultry to which the gipsies are partial. Lorca parodies the self-righteous hatred of gipsies for Civil Guards in this poem. Of all the various bodies of police instituted this century, either by Monarchists, Republicans or Nationalists, the Civil Guard were the least 'trigger-happy' of the lot. I rode among them for nine years on all the most dangerous and lonely high roads of Spain and was never so much as questioned by them. At least a half of the itinerant gipsies ('señores ambulantes' as they called themselves) were wiped out by the communists—as soon as shooting became 'free for all'—and the gipsies must have missed the Civil Guard badly in the long run. The peasants too had a feeling of hostility for the gipsies comparable to that expressed in this poem of the gipsy for the Civil Guard; and they also 'settled accounts' when they got the chance.

The Pedro Domecq, mentioned in this poem as hobnobbing with the gipsy versions of Saint Joseph and Our Lady, is the famous manufacturer of sherry at Jerez de la Frontera and almost a divinity in local gipsy eyes. Allusions to spurs and to scissors refer to the two gipsy professions of selling harness and of knife-grinding. Any gipsy quarter in Andalusia is always full of harness-shops (with assortments of out-size spurs) and grind-stones with knives and scissors, which so unheard-of and savage an irruption of the Civil Guard (as described here) would naturally scatter all over the place. A part of this poem—the fanciful description of the fair at Jerez de la Frontera—has already appeared in a preceding chapter; hence the abridgment of some lines of it here:

Their horses are as black as night
Upon whose hoofs black horseshoes clink;
Upon their cloaks, with dismal sheen,
Shine smears of wax and ink.
The reason why they cannot weep
Is that their skulls are full of lead.

With souls of patent leather
Along the roads they tread.
Hunchbacked and nocturnal,
You feel when they're at hand
Silences of india-rubber
And fears like grains of sand.
They travel where they like,
Concealing in their skulls of neuters
A blurred astronomy of pistols
And shadowy six-shooters.

When the night-time has arrived,
The night-time of the night,
Gipsy folk upon their anvils
Are forging suns and darts of light.
A wounded horse arrives and runs
To all the doors with plaintive whine.
Cocks of glass are crowing
At Jerez of the Frontier-Line.
Around the corner of surprise
The wind bursts naked on the sight,
In the night, the silver night-time,
In the night-time of the night.

The Virgin and Saint Joseph
Have left their castanets behind them
And come to ask the gipsies
If they will help to find them.
The Virgin like a Mayoress
Is sumptuously gowned
In silver chocolate paper
With almond necklets wound.
Saint Joseph moves his arms
In a silken cloak entwined
And with three Persian sultans
Pedro Domecq comes behind.
The crescent in the ecstasy
Of a white stork is dreaming
And over the flat roof-tops
Come flags and torches streaming.
Weeping before their mirrors
Hipless dancers mope and pine.
Water and shadow, shade and water
At Jerez of the Frontier-Line.

O city of the gipsies
With flags so fair to see,
Extinguish your green lamps, for here
Comes the Respectability!
O city of the gipsies
Who can forget you there?
Leave her distant from the sea
Without a comb to part her hair!

Two by two in double file
They reached the City of the Fair.
A sigh of everlasting flowers
Invades the cartridge-belts they wear.
A double nocturne of black cloth,
Their dark invasion naught deters.
Heaven to their approach appears
Merely a window-front of spurs.

The city multiplied its doors
Which, free from fear, had gaped asunder,
And through them forty Civil Guards
Enter to sack and plunder.
The clocks had stopped: the brandy
In bottles, with scared expedition,
Disguised itself with bleak November,
In order to avoid suspicion.
A flight of long-drawn screams
Ascended to the weathercock
While sabres cut the breeze with which
Their hoofs collide and shock.
The aged gipsy women fled
Along the twilight pavings,
Taking their drowsy horses
And pots filled with their savings.
Along the almost-upright streets
Sinister cloaks advance, all black
And leave a transitory vortex
Of whirling scissors in their track.

In the gateway of Bethlehem
The gipsies gather in a crowd.
Saint Joseph full of wounds,
Lays out a maiden in her shroud.
The sound of hard, sharp rifle-fire

Through all the darkness shocks and jars.
The Virgin cures the children
With the saliva of the stars.
But all the while the Civil Guard,
Advancing, sow the conflagration,
In which so tender, young, and naked,
Is roasted the imagination.
Rosa of the Camborios
Groans in a door beside the way
With her two amputated breasts
Beside her on a tray.
The other girls rush round
Chased by their flying hair
While roses of black powder
Burst round them in the air.
When all the roofs in furrows
Across the soil were strown
The morning swayed its shoulders
In a vast profile of stone.

O city of the gipsies!
The Civil Guard retires at last
Along the tunnel of the silence,
While the flames are mounting fast.

O city of the gipsies, who
That saw you could forget you soon?
Let them seek you in my forehead.
The playground of the sands and moon.

IV

The 'Canciones'

Shakespeare has defined the poet as one who gives 'to airy nothing a local habitation and a name'. If any poet could be so defined it was surely Bécquer, the exquisite nineteenth century Spanish poet who perpetually 'trembles on the verge of silence' as if he were vanishing into his own iridescence. He died young during the Romantic period and left very little poetry, but it is perhaps, in its way, the most perfect Spanish verse to be found (outside of Saint John of the Cross) in a purely disembodied form, like a rainbow. Germany is the country of romanticism and Bécquer may have derived his elfin qualities from a German grandfather, for if ever 'the horns of elf-land faintly blowing' can be heard far off in the distance, it is in the work of Bécquer, who must have cast a profound spell on Lorca.

The next collections of poems in which Lorca engaged after the *Libro de Poemas* (1921) were the two books of his songs, or *Canciones*, which were written mostly between 1921 and 1924, but not revised or published till many years later. In some of these he sets out to capture half-meanings and impressions that are difficult, vague, and remote, and literally to give to 'airy nothing' a name and address. 'Rien que la nuance', Verlaine's motto, 'nothing but the mere shade', flickers like a will-o'-the-wisp over some of these pages. Then there are enigmatical poems aiming at meanings which are beyond the usual scope of verbal combinations, as in *El Canto Quiere Ser Luz*. We have seen where previously Lorca has called the sound of water 'song made light' in *Mañana*. It is a theme that haunted him and was probably suggested by the song of Orpheus in Apollinaire's *Bestiary*, which opens:

> [Orpheus speaks]
> That signal verve of power, let it be noted!
> That majesty of line—what could be grander?
> It is the voice of light made sound as quoted
> By thrice-great Hermes in his book *Poimander*.

The old Egyptian magician, Hermes Trismegistus, was of course entirely vindicated in what he says of this phenomenon in his Hermetic books. Professor Piccard in his ascent to the stratosphere fully endorses the idea of the 'voice of light' and the 'music of the spheres', and describes how the sunbeams turned to music as they vibrated on the gondola of his balloon. This was long after Lorca wrote *El Canto Quiere Ser Luz*. I mention the fact to show that the ideas embodied in many of these obscure *canciones* are not mere shots in the dark written to bewilder or bemuse, but that they are potential ideas looking for expression. In this one case the poem was accidentally given a scientific meaning which makes it easier to consider now. But many of these ideas have the value that certain equations in the world of abstract mathematics may have for mathematicians, though they have no application or echo in the physical matter-of-fact world. Here is the poem *El Canto Quiere Ser Luz*:

> The song wishes to be light.
> In the darkness the song has

> Threads of phosphorous and moonlight.
> The light does not know what it wishes.
> Within its boundaries of opal
> It meets with itself
> And turns back home.

Here is another enigma which at least produces a visual effect of great power, *Fabula* (*Fable*):

> Unicorns and cyclopses.
>
> Horns of gold
> and eyes of green.
>
> Over the steep
> in giant confusion
> they illustrate the unglazed
> mercury of the sea.
>
> Unicorns and cyclopses.
>
> An eyeball
> and a power.
>
> Who doubts the terrible
> efficacy of those horns?
>
> Nature!
> Conceal your targets.

There are also curious nonsensical excursions into the super-real or the sub-real as in the *Canción Tonta*, (*Silly Song*):

> Mother,
> I wish to be made of silver.
>
> Son,
> You would be very cold.
>
> Mother,
> I wish to be made of water.
>
> Son,
> You would be very cold.

> Mother,
> Sew me into your cushion.
>
> O that? Yes.
> Straight away!

There comes in the second volume of the *Canciones* that uncanny and haunting *Canción de Sineta* (*Song of the Horseman*) which reminds one of the Scotch-English border ballads, which concentrate a tragedy into one or two simple verses:

> Toom came his saddle all bloody to see,
> Home came his good horse but never came he.
>   (*Bonny George Campbell*)

These border ballads have lost a good deal of their power for us today, since if any one rides out in that border country nowadays there is very little chance of his not returning quite safely; but I can remember several similar incidents on the roads around Córdoba and Sevilla, in the years of the Republic when there was all the peril expressed in this poem, especially when el Algabeño, the ex-matador and 'rejoneador' (later killed in action) was carried back unconscious with two bullets in him, from that same road. Those roads were especially dangerous; so also were the roads between Toledo, Talavera, and Ciudad Real. In June 1936 I had my partner, 'Mosquito' Bargas, shot beside me on the Talavera Road; and again in July 1936 I escaped an ambush in which I was wounded in the same district. The feeling of riding alone in the country was for many years rather fearsome during the lean years of the Republic—especially for us who depended for our living on driving livestock and horses and mules either to the abattoirs or the horse-markets. Necessitous people were often on the look-out to pick up horses or mules, or one's earnings; and many tragedies resulted. At least five people of my intimate acquaintance, all in the horse-trade, lost their lives in this way before the rising of July 17, 1936, though many years after this poem was written. I used to hate to read this poem at the time because it made me nervous about riding alone:

> Córdoba.
> Remote and lonely.
>
> Jet-black mare and full round moon,
> With olives in my saddle bags,

> Although I know the road so well
> I shall not get to Córdoba.
>
> Across the plain, across the wind,
> Jet-black mare and full red moon,
> Death is gazing down upon me,
> Down from the towers of Córdoba.
>
> Ay! The road so dark and long.
> Ay! My mare so tired yet brave.
> Death is waiting for me there
> Before I get to Córdoba.
>
> Córdoba.
> Remote and lonely.

In the consonants and the vowel sounds of this strange poem, which cannot be reproduced properly, only faintly suggested in English, one gets the rhythm of the canter of a horse, which the rider has ridden off the macadam of the main road so as to muffle the sound of his hoofs in the evening. There results a frightened, furtive, hurried, and sinister syncopation of hoofbeats and heart-beats, that defies analysis.

In the above poem Lorca is implementing the memory of popular refrains. In other poems like *El Niño Mudo* (*The Dumb Child*), he sophisticatedly attempts to give expression to ideas that have never been put into verse but belong to the dream region of 'super-realism', or call it what you like:

> The child was searching for his voice.
> (The king of the crickets had got it.)
> In a drop of water
> The child was searching for his voice.
> He did not want it for speaking with.
> I'll make a ring of it
> That he may carry my silence
> On his little finger.
>
> In a drop of water
> the child was searching for his voice.
>
> (The captive voice in the distance
> had dressed itself as a cricket.)

Throughout the *Canciones* one feels that Lorca is trying to combine the realistic matter-of-fact popular style of poems like *The Song of the Horseman* with this curious dream world of his. This he manages to do in perfect balance in the *Romancero Gitano*, and it appears, from the way he refused to publish the two books of *Canciones* until 1927 and 1932, that he regarded them in some ways as experimental note-books. Yet they are far more than that; and they also contain some very fine traditional sonnets mixed up with the more enigmatical poems. In Lorca's work, as in that of his two friends Falla and Salvador Dali, the tradition finally balances perfectly with their revolutionary innovations, though there is a violent struggle at first to dominate the irruption of the new forces in their work. Of the four Spaniards who loom so great in the world of modern art, Picasso is the only one who remains an iconoclast, in a state of perpetual fission. Here are two classical sonnets, *Sonnet* and *Adam*, from amongst the later lyrics, which show how Lorca was keeping his hand in on perfecting the classical forms, while experimenting at the same time in the haphazards of the abstruse dream world:

> Tall silver ghost, the wind of midnight sighing
> In pity opened up my ancient wound
> With his grey hand: then went and left me lying
> Where with my own sad longing I had swooned.
>
> This wound will give me life: from it will come
> Pure light and blood that issues without rest,
> A rift wherein the nightingale, now dumb,
> May find a grove, a sorrow, and a nest.
>
> O what a gentle rumour stirs my brain!
> Beside the simplest flower I'll lay my pain
> Where floats, without a soul, your beauty's pride.
>
> Then to a ruddy gold will change the vagrant
> Stream, as my blood flows out into the fragrant
> Dew-sprinkled thickets of the riverside.

And this is the second—very fine, though somewhat obscure—on Adam and the Creation of Eve:

> The morning by a tree of blood was dewed
> And near to it the newborn woman groans.
> Her voice left glass within the wound, and strewed

The window with a diagram of bones.

Meanwhile the day had reached with steady light
The limits of the fable, which evades
The tumult of the bloodstream in its flight
Towards the dim cool apple in the shades.

Adam, within the fever of the clay,
Dreams a young child comes galloping his way,
Felt in his cheeks, with double pulse of blood.

But a dark other Adam dreaming yearned
For a stone neuter moon, where no seeds bud,
In which that child of glory will be burned.

And, to show the range of the *Canciones*, I am quoting in conclusion an enigmatic little poem called *Claro de Reloj (Pause of the Clock)*:

> I sat down
> in a clearness of time.
> It was a backwater
> of silence,
> a white silence,
> wherein the stars
> went round knocking against
> black figures.

V

The Dramas

Lorca's chief public activity was concerned with the stage, in reviving the masterpieces of the Golden Age under his own direction, and in travelling round with puppet shows and companies of his own. He was responsible for a national revival of poetic drama. He wrote many plays which he produced himself. Some like *Así que pasan los Cinco Años* were experimental.

The best verse dramas of Lorca fill several volumes and are generally remarkable alike for their stage-craft and the quality of the verse that sustains them. The chief of these are briefly surveyed here.

The House of Bernarda Alba, the most powerful of them all, and the last to be written, is altogether a desolating play. It is written almost entirely in prose. All the characters are women. Bernarda, a widow with five daughters, rules her household in a small country town with a rod of iron. In an asphyxiating atmosphere of hypocrisy and tyranny, she drives all her daughters either to broken-spiritedness, abjectness, madness, or suicide. Lorca calls the three acts of it a 'documentary photograph'. It is intensely realistic. It has less imagery in it than the rest of his plays; it is a fierce satire directed at the strait-lacedness and bigotry of a certain type of matron. Nowhere in literature, for sheer loathsomeness, can we remember any character as evil as this female, Bernarda, who is a cross between a tiger and Molière's Tartuffe.

The earliest of Lorca's well-known plays is *Mariana Pineda*, a romantic historical play describing the death of a local heroine, who lived a century before, for taking part in a conspiracy against the government. The heroine is still the subject of various local ballads. Lorca's play has great dramatic qualities, and is helped out by excellent lyrical passages. The heroine, a widow, is unwilling to save her life by disclosing the other conspirators, though abandoned to her fate by one of them whom she loves, and leaving two children. She dies at the scaffold.

The next play Lorca wrote was the *Zapatera Prodigiosa, The Marvellous Shoemaker's Wife*, a light play, half-farce and half-ballet, but at the same time a searching study in female perversity. The young shoemaker's wife is a romantic type, an incurable day-dreamer, married to a hum-drum shoemaker far older than herself. She dreams of high society, makes herself unbearable, and drives out her husband, who runs away. In her imagination, during his absence, she ennobles his memory in her daydreams. He comes back in disguise, is gratified to find how she really loves and admires him, and is faithful to him in his absence. He then declares who he is; immediately she falls on him, beats him again, and the same old hell begins for both.

Yerma is a play of Spanish peasant life. *Yerma* marries a well-to-do thrifty peasant who gives her everything but children. Her childlessness and her need of children become an obsession; and though there is no lack of opportunities for having children, proffered by go-betweens, she adheres to the Spanish code of honour, but finally in a fit of exasperation kills her husband. There are some fine songs in this play, notably Yerma's song to her imaginary child, the chorus of the washerwomen at the river bank, and the parts recited by the male and female figures in the old rural masque towards the end of the play; but all these would suffer by being detached from the general context.

Doña Rosita la Soltera is a very restrained, sad play, a study in the advance of spinsterhood. The heroine is an attractive girl, pledged to a fiancé who goes abroad and finally marries someone else, though Rosita remains true to his memory. There is hardly any action in the play at all; but in the last scene Rosita's own deepening, resigned, ironic consciousness that she has had her life is magnificently brought out.

Bodas de Sangre is one of those plays that will satisfy those who think of Spain in terms of Bizet's *Carmen*. It is a story of a feud, an elopement and a mutual murder, sustained throughout by magnificent poetry with the aid of supernatural and symbolical machinery. When the whole village turns out in a 'battue' to capture and kill the eloped lovers, the hunters are aided by a Beggar Woman, whom we somehow feel to be Death, and her ally, the Rising Moon, dressed as a Woodman, who by shedding his light through the forest helps in the search for the runaways on their one horse. The *Soliloquy of the Moon*, as he joins the quest, is one of the finest passages of verse in Lorca's work:

> I am the round swan of the rivers
> And of cathedrals too the eye.
> The false dawn through the bough that shivers
> I am: from me they shall not fly!
> Who's that hiding? Who's that crying
> Down in the thickets of the vale?
> The moon has left a knife-blade lying
> Abandoned here upon the gale,
> Which, lurking hid, a glint of lead,
> Wishes to be the pain of blood.
> Let me come in! Freezing, I thread
> Through panes of glass, and walls of mud.
> To warm me, roofs and breasts divide.
> I'm freezing with the cold! The ashes
> That in my drowsy metals hide
> Seek for a crest of fire that flashes
> Through streets and mountains far and wide.
> But the snow carries me beyond
> On marble shoulders to the sky,
> And the hard waters of the pond
> Have swallowed me to freeze and die.
> Because tonight my cheeks will wear
> The tinge of blood that redly gushes
> And the broad footfalls of the air
> Will trample down the clustered rushes,

There'll be no shadow, cleft or cover
That can conceal a hunted form,
For in the broad breast of a lover
I long to enter and be warm.
Prepare a heart for me, the best,
A hot heart that will spout and flow
Through the cold mountain of my breast.
Oh, let me enter in, and glow.
I want no shadows. For my rays
Must shoot their flames through chinks and holes
And send the rumour of a blaze
Amongst the dark and mossy boles . . .

Who's hiding there? Come out and leave her,
You shan't escape among the stems.
I'll make the horse burn like a fever,
Sweating with diamonds and gems.

VI

'Cante Jondo', 'Poeta en Nueva York' and 'Llanto por Ignacio Sánchez Mejías'

The poems of Lorca's *Cante Jondo* were not published till 1928, though they were written much earlier. Here the poet seems to be writing poetry largely about the songs, dances, and festivals of Andalusia. Like all art which is founded on other art, the *Cante Jondo* therefore suffers a limitation of its appeal, especially since various poems describe songs and dances which are only known locally. However, there are a few very good lyrics, such as the following, *La Guitarra* (*The Guitar*), which have an appeal outside Andalusia:

> The lament begins
> Of the guitar.
> The wine cups of dawn
> are splintered afar.
> The lament begins
> Of the guitar.
> It's impossible, useless,
> To get it to stop.
> It weeps monotonously,
> As the rain, drop by drop,

> Or as the wind weeps
> On the snowpeak's top.
> It is impossible
> To get it to stop.
> It grieves for things
> Far out of sight—
> Like the hot southern sands
> For camellias white.
> It weeps, the targetless arrow,
> The eve without morrow,
> And the first bird on the bough
> To perish in sorrow.
> O the guitar, the heart
> That bleeds in the shades
> Terribly wounded
> By its own five blades!

Lorca went and stayed in the U.S.A. for some time, but was unable to establish a real contact with the Americans or their way of life. The result on his poetry was entirely negative. He underwent while there the intellectual influence, if not domination, of Salvador Dali, his friend, who is also a great artist of international repute, but a far more complicated personality than Lorca, more resilient and aggressive, with a far wider range of sympathies and interests, and at home anywhere from the U.S.A. to Catalonia. Lorca attempted to follow the Catalonian into the complex world of surrealism, and lost his depth. In Lorca's New York poems, the *Poeta en Nueva York*, his metaphors and images fall out of focus; his verse becomes loose, plaintive, and slightly mephitic. It took him quite a long time to recover his poetical eyesight and insight after he returned to Spain. In his long *Ode to Salvador Dali* he applauds the clear vision of the Catalan painter, but seems to lose a grasp on his own; yet the poem nevertheless contains some fine lines. We are reminded, in Lorca's American venture, of Burns when he went into high society at Edinburgh and started to write like a courtier and gentleman of the world. It was a fiasco. Lorca's talent is not cosmopolitan, and it did not flourish far from the scent of the orange groves of the South. Even in his two *Odes to the Sacrament*, which were written under the intense stimulus of religion, Lorca fails to organise the chaos in his imagery and the impact on his mind of Dali's work.

Lorca reached the height of his achievement in his *Llanto por Ignacio Sánchez Mejías*; here he remained true to his native

Andalusia, to the earth and the landscape from which his verse derived its strength, flavour and perfume; yet he was not under the restriction he imposed upon himself in the *Romancero*, that of the coldly impartial and ironic spectator. On the contrary, he was expressing his grief for a beloved friend, one of the greatest bullfighters of all time, who was also a cultured literary man, a good farmer, a great horseman, and a popular figure who was equally beloved by all for his goodness of heart, as well as for his prowess and valour. The death of Ignacio Sánchez Mejías was a public calamity in Andalusia. The matador who had retired from the bullring several years before, staged a come-back when he was well over forty. Such returns, however, are seldom successful, and it was not long before Sánchez Mejías was killed. He was the great contemporary of Joselito and Belmonte, and his name would rank after theirs, and that of the late Manolete, as perhaps the fourth-greatest matador of this century.

Whenever Lorca is inspired by the presence of bulls or horses, his verse seems to become intensified and lively, and before we go on to the *Llanto* we should remember that Lorca said when other intellectuals were talking of abolishing bullfights in Spain: 'I think it is the most cultured festival that exists anywhere in the world. It is the only place where one can go in safety to contemplate Death surrounded by the most dazzling beauty. What would happen to the Spanish springtime, to Spanish blood, and even to the Spanish language if the trumpets of the bullring should ever cease to sound?' It is rare that the fine passages of poetry which illustrate and sustain his great plays can be disengaged from their context without suffering in themselves, but there is a brilliant little picture of the bullring in Ronda given by one of the characters in *Mariana Pineda* which can be separated from the main play. From it we quote these verses:

> In the greatest bullfight ever
> At Ronda's ancient circus seen—
> Five jet-black bulls, for their devices
> Wearing rosettes of black and green [1] ...
> The girls turned up with shrilling voices
> In painted gigs and jaunting-cars
> Displaying their round fans embroidered
> With sequins glittering like stars ...
> The lads of Ronda came in riding
> Affected, supercilious mares,

[1] The badge of the terrible Miura breed.

With wide grey hats upon their eyebrows
Pulled slantwise down with rakish airs.
The tiers (all hats and towering combs)
Where people had begun to pack,
Round, like the zodiac, revolving,
Were pied with laughter white and black;
And when the mighty Cayetano
Strode over the straw-coloured sands
Dressed in his apple-coloured costume
Broidered with silk and silver bands,
From all the fighters in the ring
He stood so boldly out alone
Before the great black bulls of jet
Which Spain from her own earth had grown—
The afternoon went gipsy-coloured
Bronzing its tan to match his own.
If you had seen with what a grace
He moved his legs, and seemed to swim:
What equilibrium was his
With cape and swordcloth deft and trim:
Romero, torrying the stars
In heaven, could scarcely match with him!
He killed five bulls, five jet-black bulls
Wearing rosettes of black and green.
Upon the sharp point of his sword
Five flowers he opened to be seen.
Grazing the muzzles of the brutes,
Each instant you could see him glide,
Like a great butterfly of gold
With rosy wings fanned open wide.
The circus, with the afternoon
Vibrated, in the uproar swaying;
And in between the scent of blood
That of the mountain-tops went straying.

The *Llanto por Ignacio Sánchez Mejías* is divided into four parts. The first part harrows us with its hysterical refrain 'at five in the afternoon', as it describes the commotion when the matador is tossed; with the medical preparations, the smell of iodine in the infirmary, the strewing of chloride on the sand; it reaches the side of the dying matador in the following verses:

A coffin on wheels is the bed
At five in the afternoon.

Bones and flutes sound in his ears
At five in the afternoon.
The bull was bellowing through his forehead
At five in the afternoon.
The room was rainbowed with agony
At five in the afternoon.
From far away the gangrene comes already
At five in the afternoon.
The trumpet of the lily through green groins
At five in the afternoon.
Like suns his wounds were burning
At five in the afternoon.
And the crowd was breaking the windows
At five in the afternoon.
At five in the afternoon.
Ay, what a terrible five in the afternoon!
It was five by all the clocks!
It was five in the shade of the afternoon.

Then follows the second movement entitled *The Spilt Blood*:

I do not want to look at it!

Tell the moon it's time to rise,
I do not want to see his blood
Where spilt upon the sand it lies.

I do not want to look at it!

The moon in open spaces lit,
Horse of the quiet clouds, is showing,
And the grey bullring of a dream
With willows in the barriers growing.
I do not want to look at it!
Let my remembrance burn away.
Inform the jasmine-flowers of it
Within their tiny stars of spray!

I do not want to look at it!
The cow of this old world was licking,
With its sad tongue, a muzzle red
With all the blood that on the sand
Of the arena had been shed.
And the five bulls of Guisando

Half made of death, and half of granite,
Like centuries began to low,
Grown tired of trampling on this planet;
No.
I do not want to look at it!

With all his death borne on his shoulders
Ignacio ascends the tiers.
He was looking for the daybreak
Where never break of day appears.
He sought for his accustomed profile,
But the dream baffled him instead.
He looked to find his handsome body
But found his blood was opened red.
Don't ask of me to look at it!
I do not wish to smell the source
That pumps each moment with less force,
The stream by which the tiers are lit,
The stream that spills its crimson course
Over the corduroy and leather
Of the huge crowds that thirsting sit.
Who shouts to me to have a look?
Don't tell me I should look at it!

He did not try to close his eyes
When he saw the horns so nigh,
But the terrible mothers
Lifted up their heads on high.
And through the ranching lands a wind
Of secret voices started sighing
That to the azure bulls of heaven
Pale cowboys of the mist were crying.
In all Seville to match with him
Has never lived a prince so royal,
Nor any sword to match with his,
Nor any heart so staunch and loyal.
Like a torrent of lions, his
Incomparable strength was rolled.
And like a torso hewn in marble
His prudence carven and controlled.
Gold airs of Andalusian Rome
Circled his head and gilded it,
Whereon his laugh was like a lily
Of clear intelligence and wit.

How great a fighter in the ring!
How good a peasant in the shire!
How gentle with the ears of corn!
And, with the spurs, how hard and dire!
How soft and tender with the dew!
How bright our fair-days to illume!
How tremendous with the final
Banderillas of the gloom!

But now he sleeps without an end,
Now the world masses and the grass,
Opening the lily of his skull,
Their fingers may securely pass.
And now his blood comes singing as it flows,
Singing by swamps and fields beyond control,
Gliding around the stiff horns of the snows
And wavering in the mist without a soul
Like a long, dark, sad tongue it seems to slide
Meeting a thousand cloven hoofs, and flies
To form a pool of agony beside
The starry Guadalquivir of the skies.
O white wall of Spain!
O black bull of pain!
O hard blood of Ignacio!
O nightingale of his red vein!
No!
I do not want to look at it.
There is no cup to hold it fit.
There are no swallows fit to light on it,
No frost of light is fit to whiten it,
No song, nor shower of lilies over it,
Nor glass with silver screen to cover it.
No!
I will not look at it.

The third movement describes the wake of the matador's friends by his body as it lies in state, and ends impressively with the following stanzas:

Iwish none here but those whose voices ring.
Tamers of steeds and rivers, from the heath.
The men whose skeletons resound, who sing
With flint and sunlight flashing from their teeth.

> I want to see them here. Before this stone—
> This body with its snapped and trailing reins.
> By them I want to see the exit shown
> For this great Captain whom death leads in chains.
>
> And like great rivers let them teach me dirges
> That have soft mists and canyons steep and full
> To bear Ignacio's body on their surges
> Far from the double snort of any bull:
>
> Till, lost in the arena of the moon
> That feigns itself a bull in a childish play,
> They leave him, where no mortal fish may croon,
> In the white thickets of the frozen spray.
>
> I would not have them hide his face with cloth
> To wean him to the death in which he lies.
> Ignacio, go! Though bellowing bulls may froth—
> Sleep, fly, and rest. Even the ocean dies!

The fourth part of the *Llanto* in which he takes final leave of his friend, ends with a verse which might serve as an epitaph for the poet himself:

> It will be long before there is born, if ever,
> An Andalusian so frank, so rich in adventure;
> I sing your elegance with words that moan
> And remember a sad wind among the olive trees.

BIOGRAPHICAL NOTE

Federico García Lorca was born at Fuente Vaqueros near Granada on January 5, 1899.

He studied at Almería, a nearby sea-port, and later at Granada University, where he became a licentiate in law. At Granada University he began a course in Philosophy and Letters, which, in 1919, he continued at Madrid University, where his fame as a poet was instantaneous though he did not publish till two years later. His poetry became known through recitation. At the University of Madrid he met the painter Salvador Dali, the poet Gerardo Diego and many other leading spirits of his generation. He was a gifted pianist and painter besides being a poet, actor and dramatist. His *Libro de Poemas* was published in 1921 and his first dramatic success was scored with *Mariana Pineda* in Barcelona in 1927. He held a successful show of pictures there in the same year. In 1928 along with de Falla and other brilliant poets and musicians of Granada he founded and edited *Gallo*. In 1929 and 1930 he travelled through the United States, Canada and Cuba. In 1933 and 1934 he produced classical Spanish plays with a company of his own in Buenos Aires and Monte Video. Amongst other plays he produced his own with great success and gave many lectures on literature and folk-lore while he was abroad. In 1934, he settled in Madrid, devoting himself chiefly to dramatic work. On the outbreak of the terror in Madrid he fled to Granada where he was murdered in July 1936.

LORCA'S PUBLISHED WORK

Impresiones y Paisajes, Ventura, Granada, 1918, (prose).
Libro de Poemas, Imprenta Maroto, Madrid, 1921.
Canciones, 2nd. ed., Revista de Occidente, Madrid, 1929.
Primer Romancero Gitano, Revista de Occidente, Madrid, 1928.
Poema del Cante Jondo, C.I.A.P., Madrid, 1931.
Oda a Walt Whitman, Alcancía, Mexico, privately printed, 1933.
Llanto por Ignacio Sánchez Mejías, Cruz y Raya, Madrid, 1935.
Obras Completas, Editorial Losada, Buenos Aires, 1938 (new edition 1946).

PLAYS

Mariana Pineda, La Farsa, Madrid, 1928.
Bodas de Sangre, Cruz y Raya, Madrid, 1936.
La Casa de Bernarda Alba, Editorial Losada, Buenos Aires, 1945.

The *Obras Completas* contains nearly all Lorca's poems and plays except for the important play *La Casa de Bernarda Alba* which the same publisher has issued in a volume by itself. A few unpublished poems and fragments of prose appeared at the back of *Federico García Lorca*, the monumental critical appreciation by Guillermo Díaz-Plaja, published by Guillermo Kraft Limitada, Buenos Aires.

DATE DUE

DEC 14 '67			
MAY 11 '68			
MAY 26 '69			
SEP 29 1994			

861 25867
G16yc
Campbell
AUTHOR

Lorca
TITLE

861 25867
G16yc